Hybrid Nanostructures as Solid-State Sensors for IoT

This book provides an in-depth discussion of various hybrid nanostructures as solid-state sensors in the context of the Internet of Things (IoT). It explains the vital role of sensors in IoT to aid in discovering what possibilities ought to be addressed to make the data more meaningful. The book highlights the applicability of biosensing field-effect transistor (FET) technology with a specific emphasis on the progress being made in integrating existing FET technology and nanotechnology, using semiconductor nanowires and organic structures.

Features

- Offers in-depth discussion on hybrid nanostructures as solid-state sensors
- Elaborates upon the fabrication of gas sensors using metal oxide semiconductor nanostructures
- Studies biosensors based on metal oxide semiconductor nanostructures
- Discusses the suitability of various hybrid sensors in solving the problems of the IoT
- Provides extensive support for the development of hybrid nanostructures as solid-state sensors

This book is meant for researchers and scholars of computer science and associated disciplines. It also serves as a valuable reference for graduate students and researchers seeking to deepen their knowledge and engage with the latest advancements in these hybrid nanostructures as solid-state sensors.

Hybrid Nanostructures as Solid-State Sensors for IoT

Edited by
Aniruddha Mondal and Arindam Biswas

CRC Press is an imprint of the
Taylor & Francis Group, an **informa** business
A CHAPMAN & HALL BOOK

First edition published 2025
by CRC Press
2385 NW Executive Center Drive, Suite 320, Boca Raton FL 33431

and by CRC Press
4 Park Square, Milton Park, Abingdon, Oxon, OX14 4RN

CRC Press is an imprint of Taylor & Francis Group, LLC

ISBN: 9781032714035 (hbk)
ISBN: 9781032734439 (pbk)
ISBN: 9781003464211 (ebk)

DOI: 10.1201/9781003464211

Typeset in Minion
by Newgen Publishing UK

Contents

List of Abbreviations, vii

Preface, xi

Acknowledgements, xiii

About the Editors, xv

Contributors, xvii

Chapter 1 ▪ Introduction 1

Naorem Khelchand Singh

Chapter 2 ▪ Preparation of Metal Oxide Semiconductor Nanostructure and Their Structural Properties 16

Paulsamy Chinnamuthu, Iman Biswas, and Aniruddha Mondal

Chapter 3 ▪ Polymer Semiconductors and FET Sensors 39

Alexy R. Tameev

Chapter 4 ▪ Fabrication of Gas Sensor Using Metal Oxide Semiconductor Nanostructures 59

Bijit Choudhuri and Parul Raturi

CHAPTER 5 ▪ Study of Biosensors Based on Metal Oxide Semiconductor Nanostructures 121

PRIYANKA CHETRI, JAY CHANDRA DHAR, ASHOK PALEPU, AND MICHAEL CHOLINES PEDAPUDI

CHAPTER 6 ▪ Biosensor Overview and Its Advances for Cancer Detection 150

PRASANNA KARKI, BIBEK CHETTRI, PRONITA CHETTRI, SANAT KR. DAS, AND BIKASH SHARMA

CHAPTER 7 ▪ Recent Progress in Nanomaterial-Based IoT-Enabled Sensor 166

RAHUL SAMANTA, SANDIP KUNAR, GURUDAS MANDAL, ATUL BANDYOPADHYAY, ARINDAM BISWAS, AND SWARUP KUMAR GHOSH

INDEX, 187

List of Abbreviations

0D	zero-dimensional
1D	one-dimensional
2D	two-dimensional
3D	three-dimensional
ADC	analog-to-digital converter
BWA	biowarfare agents
CMOS	complementary metal-oxide semiconductor
CNT	carbon nanotube
CQDs	carbon quantum dots
CPM	constant photocurrent method
CWEs	coated-wire electrodes
CVD	chemical vapour deposition
DBT	dithienyl-2,1,3-benzothiadiazole
DLS	dynamic light scattering
DNA	deoxyribonucleic acid
DoS	density of states
DPP	diketopyrrolopyrrole
EDAX	energy-dispersive X-ray analysis
EDL	electron depletion layer
EIS	electrochemical impedance spectroscopy
FACS	fluorescence-activated cell sorting
FET	field-effect transistor
FRET	Förster resonance energy transfer
FTO	fluorine-doped tin oxide
FWHM	full-width at half-maximum
GCE	glassy carbon electrode
GFP	green fluorescent protein
GLAD	glancing angle deposition

GQDs	graphene quantum dots
HAL	hole accumulation layer
HAS	human serum albumin
Hb	hemoglobin
HDAC	histone deacetylase
HOMO	highest occupied molecular orbital
IoT	Internet of Things
ISE	ion selective electrode
ISFETs	ion selective field-effect transistor
ITO	indium tin oxide
IVDS	in-vitro diagnostics
LDM	low-dimensional materials
LEDs	light emitting diodes
LPE	liquid phase epitaxy
LUMO	Lowest Unoccupied Molecular Orbital
MB	molecular beam
MBE	molecular beam epitaxy
MBV	mean boundary velocity
MNPs	magnetic nanoparticles
MOCVD	metal-organic chemical vapour deposition
MOFs	metal-organic frameworks
MONP	metal-oxide-based nanoparticles
MWCNT	multi-walled carbon nanotube
NPs	nanoparticles
NTCDIs	naphthalene tetracarboxylic diamides
NWs	nanowires
OAD	oblique angle deposition
OFET	organic field-effect transistor
OSCs	organic and/or polymer semiconductors
PA	Polyacetylene
PCB	polymer circuit board
PECVD	plasma enhanced chemical vapor deposition
P3HT	poly(3-hexylthiophene)
POC	point of care
POCT	point of care testing
POS	p-type oxide semiconductors
PSMA	prostate-specific membrane antigen
PVC	polyvinyl chloride

QDs	quantum dots
RNA	ribonucleic acid
SAM	self-assembled monolayer
SMO	semiconducting metal oxide
SMU	source measure unit
SVR	surface-to-volume ratio
SWCNT	single-walled carbon nanotube
TC	transport centers
TF	thin film
TLV	threshold limit value
TMDs	transition metal dichalcogenides
TPA	triphenylamine
UTI	urinary tract infection
UV	ultraviolet
VLS	vapour-liquid-solid
VOC	volatile organic compound
VS	vapour solid

Preface

The Internet of Things (IoT) is reshaping our world, creating a network of devices that communicate to enhance our daily lives. Central to this transformation are advanced sensors that detect and respond to various stimuli. *Hybrid Nanostructures as Solid-State Sensors for IoT* explores the promising field of hybrid nanostructures, which offer exceptional sensitivity and specificity for IoT applications.

Edited by Dr Aniruddha Mondal and Dr Arindam Biswas, this book provides a detailed examination of the design, fabrication, and application of these sensors. The chapters offer both foundational knowledge and the latest research, making it a vital resource for students, academics, and professionals.

Chapter 1 introduces the fundamental concepts and significance of hybrid nanostructures. Chapter 2 covers the synthesis and structural properties of metal oxide semiconductors. Chapter 3 explores polymer semiconductors and field-effect transistor (FET) sensors. Chapter 4 discusses the fabrication of gas sensors using metal oxide nanostructures. Chapter 5 delves into biosensors based on metal oxide semiconductors, and Chapter 6 reviews advances in biosensors for cancer detection. Chapter 7 includes an extensive examination summarizing the latest worldwide advancements in research concerning the evolution of Internet of Things (IoT)-enabled sensor materials.

We thank all contributing authors for their dedication and insights, which have enriched this volume. We hope this book serves as a valuable resource and inspiration for further innovations in sensor technology and IoT applications.

Dr Aniruddha Mondal
Dr Arindam Biswas

Preface

Acknowledgements

THE AUTHOR NAOREM KHELCHAND Singh acknowledges the National Institute of Technology Nagaland, India, for Chapter 1 of this work. The authors Iman Biswas, Aniruddha Mondal, and Paulsamy Chinnamuthu acknowledge the National Institute of Technology Durgapur, India, and the National Institute of Technology Nagaland, India, for Chapter 2. The author Alexy R. Tameev acknowledges Frumkin Institute of Physical Chemistry and Electrochemistry, Russian Academy of Science (ICPE RAS), Russia, for Chapter 3. The authors Parul Raturi and Bijit Choudhuri acknowledge Omkaranand Saraswati Government Degree College, Uttarakhand, India, and the National Institute of Technology Silchar, India, for Chapter 4. Priyanka Chetri, Jay Chandra Dhar, Ashok Palepu, and Michael Cholines Pedapudi acknowledge the Department of Electronics and Communication Engineering, National Institute of Technology Nagaland, India, for Chapter 5. Also, one part of this work was supported by the All India Council for Technical Education (AICTE), Government of India, under the Research Promotion Scheme for North-East Region (RPS-NER) vide ref.: File No. 8-139/RIFD/RPS-NER/Policy-1/2018-19 (Principal Investigator: Bikash Sharma). The author Sanat Kr. Das acknowledges the TMA Pai University Research Fund Award for Minor Grant, Sikkim Manipal Institute of Technology, Sikkim Manipal University, Sikkim (Sanction No.: 6100/SMIT/R&D/Project/12/2020, dated 20th July 2020). The authors Rahul Samanta, Gurudas Mandal, and Arindam Biswas acknowledge Kazi Nazrul University, West Bengal, India; the author Sandip Kunar acknowledges Aditya Engineering College, Andhra Pradesh, India; the author Atul Bandyopadhyay acknowledges the University of Gour Banga, West Bengal, India; and the author Swarup Kumar Ghosh acknowledges the Indian Institute of Engineering Science and Technology, West Bengal, India, for Chapter 7.

About the Editors

Aniruddha Mondal has been an Associate Professor in the Department of Physics at the National Institute of Technology Durgapur since 2014. He earned his PhD in Electronic Science from Calcutta University in 2008. He then served as Institute Postdoctoral Fellow in the Department of Electrical Engineering at the Indian Institute of Technology Kanpur (2008–2009). Then, he was appointed as Research Professor/Assistant Professor at Dongguk University in Seoul, Republic of Korea (2009–2010), where he worked at the Millimeter Wave Innovation Technology Research Centre. He also worked as Assistant Professor at the National Institute of Technology Agartala, Tripura, in the Department of Electronics and Communication Engineering for 4.5 years (2010–2014). His research interests include the fabrication of III-V, III-N, perovskite, and thermochromic thin films and nanostructure semiconductor materials, 1D metal oxide semiconductor nanostructure using glancing angle deposition technique, and fabrication of UV-Vis detector, infrared detector, plasmonic detector, and hybrid semiconductor detectors. He has published more than 88 papers in highly reputed Science Citation Index (SCI) journals, 58 conference papers, and holds 3 patents. He has guided 13 PhD students, and 3 PhD students are currently working under his supervision along with one Postdoctoral Fellow/RA. He has completed seven projects as PI, funded by the Department of Science and Technology (Science and Engineering Research Board, India) [DST (SERB)], Board of Research in Nuclear Sciences (Department of Atomic Energy, India) [BRNS (DAE)], All-India Council for Technical Education (AICTE), and Indian

Space Research Organization (ISRO). Currently, two projects funded by DST (SERB) and ISRO are running under his supervision.

Arindam Biswas is presently an Associate Professor in the School of Mines and Metallurgy at Kazi Nazrul University, West Bengal, India. He also serves as adjunct faculty at Taylor's University, Malaysia. He earned an MTech degree in Radio Physics and Electronics from the University of Calcutta, India, in 2010 and completed his PhD at NIT Durgapur in 2013. He has worked as a post-doctoral researcher at Pusan National University, South Korea, with the prestigious BK21PLUS Fellowship, Republic of Korea. Biswas has received the DST-JSPS Invitation Fellowship at RIE, Japan, and the DST-ASEAN Invitation Fellowship at Duy Tan University, Vietnam, and Taylor's University, Malaysia. He has also served as Visiting Fellow in the Department of Electrical and Computer Engineering at the National University of Singapore. He has worked as Specially Appointed Associate Professor (Visiting) at the Research Institute of Electronics, Shizuoka University, Japan. He has been selected for IE (I) Young Engineer Award for 2019–20, KNU Best Researcher Award (Engineering and Technology) for 2021, and KNU Best Faculty Award for 2022 (Faculty of Science). He has authored 8 books/monographs and edited 24 volumes and 10 book chapters with international repute. Biswas has received research grants from various funding agencies, including SERB, DST-ASEAN, DST-JSPS, UGC, and industrial consultancy from Eastern Coalfield Limited (ECL). He has also received an international research grant from the Centre of Biomedical Engineering, Tokyo Medical and Dental University, in association with RIE, Shizuoka University, Japan, for four consecutive years from 2019 to 2023. Biswas has guided six PhD students, with an additional three students having submitted their theses for PhD. He is a life time member of the Institute of Engineers (India), the Mining, Geological & Metallurgical Institute of India (MGMI), Regular Fellow of Optical Society of India, and a member of the Institute of Electrical and Electronics Engineers (IEEE).

Contributors

Atul Bandyopadhyay
University of Gour Banga
West Bengal, India

Arindam Biswas
Kazi Nazrul University
West Bengal, India

Iman Biswas
National Institute of Technology Durgapur
West Bengal, India

Bibek Chettri
Sikkim Manipal Institute of Technology
Sikkim, India

Priyanka Chetri
National Institute of Technology Nagaland
Nagaland, India

Pronita Chettri
Sikkim Manipal Institute of Technology
Sikkim, India

Paulsamy Chinnamuthu
National Institute of Technology Nagaland
Nagaland, India

Bijit Choudhuri
National Institute of Technology Silchar
Assam, India

Sanat Kr. Das
Sikkim Manipal Institute of Technology
Sikkim, India

Jay Chandra Dhar
National Institute of Technology Nagaland
Nagaland, India

Swarup Kumar Ghosh
Indian Institute of Engineering Science and Technology
West Bengal, India

Prasanna Karki
Sikkim Manipal Institute of Technology
Sikkim, India

Sandip Kunar
Aditya Engineering College
Andhra Pradesh, India

Gurudas Mandal
Kazi Nazrul University
West Bengal, India

Aniruddha Mondal
National Institute of Technology Durgapur
West Bengal, India

Ashok Palepu
National Institute of Technology Nagaland
Nagaland, India

Michael Cholines Pedapudi
National Institute of Technology Nagaland
Nagaland, India

Parul Raturi
Omkaranand Saraswati Government Degree College
Uttarakhand, India

Rahul Samanta
Kazi Nazrul University
West Bengal, India

Bikash Sharma
Sikkim Manipal Institute of Technology
Sikkim, India

Naorem Khelchand Singh
National Institute of Technology Nagaland
Nagaland, India

Alexy R. Tameev
Frumkin Institute of Physical Chemistry and Electrochemistry
Moscow, Russia

CHAPTER 1

Introduction

Naorem Khelchand Singh

1.1 NEED FOR SENSORS ON THE INTERNET OF THINGS AND POINT-OF-CARE DEVICES

A sensor is primarily a tool that can detect specific changes in an environment or communicate with a physical environment. It is an instrument capable of measuring physical parameters, such as temperature, pressure, humidity and so on, and converting them into an electric signal. Every good sensor should have the following three traits:

- It should be sensitive to the measured phenomenon.
- It should not be prone to any other physical phenomena.
- The measured phenomenon should not be changed during the measurement process.

There are more than 20 million devices such as smartphones, wearables, and other devices all around us which employ sensors in one way or the other. We can use a wide variety of sensors to quantify virtually all the physical properties around us. Various types of sensors are used in several Internet of Things (IoT) applications [1–2]. The smartphone is the most standard and common sensor currently available. There are several types of sensors installed in the smartphone itself [3] such as the position sensor (GPS), motion sensors (accelerometer, gyroscope), camera, light sensor, microphone, proximity sensor, and magnetometer. These are extensively used in various IoT applications. Other sensors that are commonly implemented in day-to-day life usually involve thermometers, pressure

DOI: 10.1201/9781003464211-1

sensors, light sensors, motion sensors, gas sensors, body parameter sensors such as heartbeat, blood sugar levels, chemical and biochemical substances, neural signals, and several others. Infrared sensors that precede smartphones belong to the category of sensors that prove exceptional. They have become commonly useful in numerous IoT applications: IR cameras, motion detectors, remote measurement of nearby objects, presence of smoke and gases, and sometimes as humidity sensors. Such sensors play a significant role in the monitoring of air pollution levels, state of health, home security, and industrial IoT (IIoT) manufacturing processes.

The two vital aspects of an IoT framework are the Internet and hardware components like sensors and actuators. In any IoT architecture such as the one given in Figure 1.1 [4], the ground layer of the IoT framework consists of sensor communication and network for gathering information. Sensors thus have the basic function of sensing and gathering information from the surroundings, and thus sensors are the motors of IoT. Sensors, or "things" of the IoT system, shape the front end. It detects certain physical parameters in the environment or recognizes certain smart objects.

The sensor array of data must be stored and analysed in an intelligent manner so that critical data can be extracted from it. After signal transformation and processing, these are linked implicitly or explicitly to IoT networks. Since all sensors are distinct, different IoT implementations need different sensor kinds. Digital sensors, for example, are simple and easy to communicate with a microcontroller using a Serial Peripheral Interface (SPI) bus. On the contrary, either analogue-to-digital converter (ADC) or Sigma-Delta modulator is used for converting the data to SPI output in

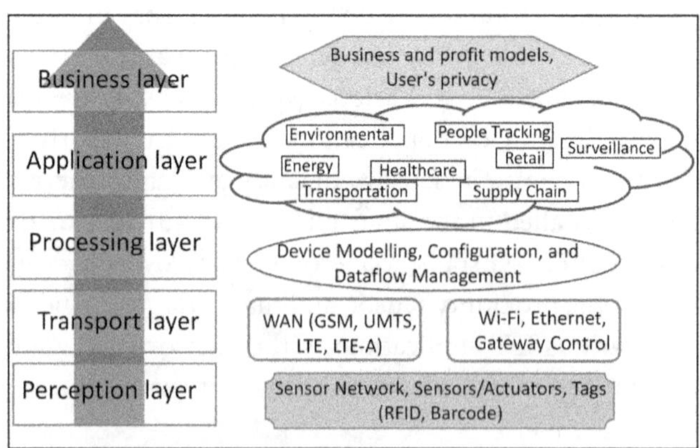

FIGURE 1.1 IoT architecture layers.

the case of analogue sensors. Usually, the data generated are sent through a remote server. An IoT object's storage and processing capacities are often limited by the available resources, which are often much curtailed due to the constraints in size, space, power, and computation powers. The key research concern as a natural consequence is to guarantee that we get the appropriate amount of information at a high degree of precision from the sensors.

The IoT has transformed the world, indeed. It affects our way of life, as well as our activities. Currently, with broadband Internet accessible and reasonably priced anywhere around, and Wi-Fi technologies and sensors integrated into a wide range of devices and gadgets, wearable and smartphone penetration is booming. All these things have paved the way for the IoT in daily lives and in other sectors of the economy, including healthcare. Healthcare IoT has become a sector with many potentialities, and it incorporates digital health technologies to increase access to medical care and reduce overall costs. In addition to the pervasive smartphones, three significant technological developments converge to sustain this modern healthcare era of advanced medical diagnostics [5]: (1) biological data collection by biosensors, (2) diagnostic point of care (POC), and (3) IoT.

In this context and with reference to Figure 1.2, IoT alludes to wearable or mobile gadgets like smart watches, pulse indicators, and sugar indicators equipped with circuitry, programs, sensors (including biosensors), and network access that enables data to be obtained and exchanged. POC devices are used to conduct diagnostic tests and acquire results by medical

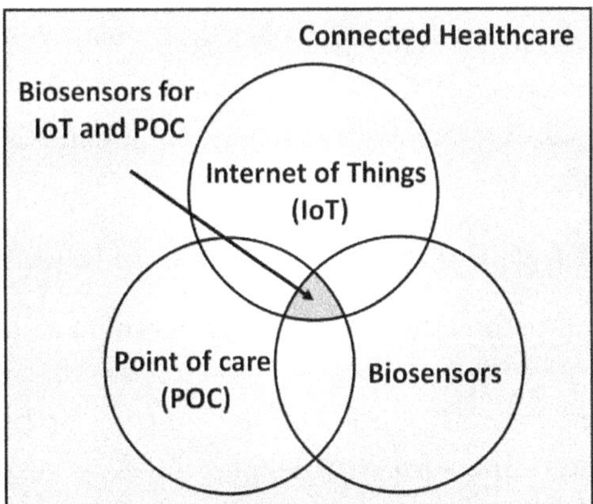

FIGURE 1.2 Relationship between biosensors, point of care (POC) testing, and Internet of Things (IoT).

professionals when being with or near the patient within a short span of time and to allow medical self-testing at home. Such devices are typically biosensors in the form of wearable or portable external devices that track physiological data with remote/wireless communication that is used for telemedicine and inpatient monitoring. These devices, for example, monitor blood pressure, ECG, temperature, periodic glucose, level of oxygen, and so on.

Point of care testing (POCT) devices can be categorized into two types: analysers (in vitro sampling) or monitors (ex vivo, in viva sampling). POCT analysers are most efficient in ambulatory and emergency units, while monitoring devices are more functional in the environment of intensive care treatment. Analysers and monitors both use biosensor technology to detect analytes. Biosensors can be either electrochemical or optical sensors. The majority of POCT analysers employed nowadays use electrochemical sensors. Research and development of new biosensors based on fluorescence technology [6] and immunochromatographic assays [7] is underway.

POC-based in vitro diagnostics (IVDs) are gaining recognition because healthcare personnel and facilities are not necessary to deliver results [8]. Samples like blood, saliva, urine, and other body fluids do not need pre-preparation (or can necessitate a minimal amount of pre-preparation). The tests yield results within seconds in several scenarios. On the contrary, the processing time of normal test results is in hours or even days in a normal hospital, clinic, or office of a physician. The assessments of POC test results are easy to perform and easy to interpret the findings. Recently, POC devices are beginning to move from single-use test samples to paper strips or colours to compact or hand-held electronic equipment that produce progressively detailed reports. POC devices are significantly relying on new methodologies including molecular imaging for identification and prevention, disease or disorder diagnosis, and treatment [9].

1.2 BASIC DEFINITIONS RELATED TO SENSING

Sensor efficiency is defined by the following parameters:

- **Sensitivity:** Most of the sensors have a transfer function which is linear in nature. The sensitivity is then defined as the ratio of the output signal to the measured property. Sensitivity is simply the smallest fractional shift detectable in a system. Highly sensitive sensors experience greater output variability because of input variability.

- **Selectivity:** Selectivity is the ability of a sensor to determine the concentration of a substance in comparison to other interfering agents. Selectivity refers to characteristics that decide whether a sensor is responsive to a group of analytes selectively or even to a single analyte in specific. It is used to characterize the capability of the sensor system to detect certain components of an unknown mixture (qualitative analysis).
- **Reversibility:** Reversibility is the sensor's ability to recover after being exposed to an analyte or bounce back to its original background/ baseline state.

1.3 OVERVIEW OF SENSING TECHNIQUE

Almost all medical devices used in the laboratory as well as at the POC employ sensors to measure analytes such as electrolytes, pH, blood gases, and metabolites. Electrochemical methods perhaps belong to the earliest methods of measurement that consist of a wide variety of data analysis possibilities, augmented by modern technical advances in signal analysis and digital systems. The basics of the existing electrochemical measuring approaches used in specific sensors are dealt with below by paying a brief glance at some latest techniques as well.

1.3.1 THE ELECTROCHEMISTRY APPROACHES

Electrochemical process-based sensors have been used in conventional in vitro diagnostic techniques for determining pH, gasses, and electrolytes in the whole blood for years [10]. It works on the concept that an electrical current passes across a sensing electrode generated by an electrochemical reaction that occurs on the surface of a catalyst-coated sensing electrode, such as platinum [11]. They are particularly recognized due to their excellent delectability, easy operation, and cost effectiveness. They have been proven to be the preferred choice for the monitoring of critical care analytes, for POCT, attributing to the fact that they can be utilized in a turbid and complicated sample matrix such as whole blood in a direct manner for rapid study. They are reversible and very much flexible as compared to other sensors such as optical sensors for multipurpose applications as they depend on interfacial instead of bulk chemical sensing reactions. They find a prominent place among the sensors available in the market today which has found a wide variety of potential applications in the clinical, manufacturing, environmental, and agricultural research fields. These kinds of sensors use a variety of detection mechanisms, such as potentiometry, amperometry, conductimetry, and impedimetry, depending on the specific mode of signal transduction. The fundamentals of electrochemical methods used in most POCT systems are briefly discussed here. These approaches are used to quantify many items of clinical interest directly and to create platforms for biosensors.

1.3.2 Potentiometry

Due to their simplicity, familiarity, and low cost, potentiometric sensors have been widely used since the early 1930s. Potentiometry is based on the formation of voltage difference between two electrodes (half cells) with the cell current almost negligible. The sample connects the two electrodes, and they are labelled as the indicator and reference electrodes. The reference electrode is needed to supply a half-cell potential constantly. The indicator electrode produces a varying potential based on the activity or distribution of a particular solution component. A high-input impedance voltmeter (input impedance $>10^{12}\ \Omega$) is used to calculate the electrical potential difference between the electrodes and is needed to sustain zero current flow in the cell to yield precise potential measurements.

The total cell potential is associated with the Nernst equation's ion of interest activity in the sample [12]:

$$E_{cell} = E^o + \left(2.303\frac{RT}{zF}\right)\log a_I \tag{1.1}$$

where E^o denotes the standard cell potential, R denotes the molar gas constant, T indicates the temperature (K), z denotes the charge of the ion of interest, F is the Faraday constant, and a_I is the activity of the ion of interest.

Three specific types of potentiometric devices exist: ion-selective electrodes (ISE), coated wire electrodes (CWES), and field effect transistors (FETS). ISE is by far the most popular potentiometric sensor [13]. The ISE employs an indicator electrode which determines the activity of a specific analyte ion specifically. The ion-selective membrane is structured to provide a potentially selective signal for the target ion and is positioned at the electrode tip. Because of the preferential separation of the ionic species between these two levels, this potential signal is produced at the interface between the ion-selective membrane and solution due to charge separation.

The outcome is evaluated under generally zero current conditions. Speedy and reversible response of the indicator electrode is preferred and should be driven by the Nernst equation.

The ISEs can be classified into three categories, based on the type of membrane: glass, liquid, or solid electrodes [14]. The most widely recognized glass electrodes are centred on a thin ion-sensitive glass membrane and are obtainable in various forms. Glass electrode applications have also been recorded for monovalent cations such as sodium [15], lithium [16], ammonium, and potassium sensors [17] obtained from new glass compositions. These electrodes are restricted in use and have conventionally only been around for pH and Na^+ analyses. For explicit measurements of numerous polyvalent cations and some other anions, liquid-membrane-electrode ISE reliant on water-insoluble liquid substances infused with a polymeric membrane is utilized. The polymeric membrane is employed to segregate the sample solution from the internal compartment which contains the target ion solution. A liquid ion exchanger or a neutral macrocyclic compound comprising voids to envelope the target ions could be used for membrane-active identification [18, 14].

Membranes of polyvinyl chloride (PVC) are usually used for K^+, Na^+, Cl^-, Ca^{++}, Li^+, Mg^{++}, and pH analyses. Despite not being as robust as glass membrane electrodes, the PVC membrane electrodes are far more flexible. A standard proportion of an ISE based on PVC membrane is characterized as 1–3 w/w% neutral ion exchanger, ~64 w/w% plasticizer, ~30 w/w% PVC, and <1 w/w% additives. The critical element of the membrane in assessing the selectivity of one ion against others is the charged ion exchanger also known as an ionophore. The finding that the antibiotic valinomycin was being integrated into a plasticized PVC membrane and can act as an ionophore with high selectivity for K^+ rather than Na+ is an integral step in the progress and routine application of PVC type ISEs [19]. The valinomycin-based K^+ ISE was the first representation of an ISE-neutral carrier and is commonly adopted nowadays for daily blood K^+ evaluation.

Freiser first developed coated-wire electrodes (CWEs) in the mid-1970s [20]. In the traditional CWE model, a conductor is directly layered with a suitable ion-selective polymer membrane such as PVC, poly (vinyl benzyl chloride), or poly (acrylic acid) to develop an electrolytic concentration-sensitive electrode. Concerning detectability and concentration rate, the CWE outcome is close to that of conventional ISE [20]. This setup removes the necessity of an internal reference electrode and proves to be a major advantage and thus culminate in outcomes which is beneficial while miniaturization. It is extremely effective for the biomedical and clinical investigations of various kinds of analytes for in vitro and in vivo models [21, 22].

ISEs are integrated with the semiconductor field effect transistor (FET) model to develop ion-selective FETs (ISFETs) where an ion-selective membrane has substituted the gate. The FET, a solid-state device with high input impedance and low output impedance, can control charge building up on the membrane. ISFETs having bare gate insulators such as silicon oxide, silicon nitride, and aluminium oxide demonstrate inherent pH-sensitivity because of electrochemical equilibrium between the protonated oxide surface and the solution protons [23]. ISFETs have the advantages of small size and sturdy design, making it an appropriate sensing tool for environmental and industrial research. More about FET-based sensors are discussed in the following sections.

1.3.3 Amperometry

When powered by a continuous, applied voltage potential, some chemical species referred to as electroactive substances can be oxidized or reduced at inert metal electrodes using electrochemical cells. The substance is reduced to gain electrons in a reduction reaction; in an oxidation reaction, an oxidizing substance loses electrons. The anode and cathode are the electrodes of the electrochemical cell at which oxidation and reduction take place. The key component for the basic form of amperometric biosensors may be the Clark oxygen electrodes, where the oxygen content is proportionate to the production of current. It is done by reducing oxygen in a platinum-operated electrode in comparison to an Ag/AgCl reference electrode provided that a potential exists [24]. The current is primarily determined when the potential is supplied constantly, and this is called amperometry. When the current is calculated during regulated potential variations this is called voltammetry. Since many protein analytes are naturally incapable of operating as redox partners in electrochemical reactions, these tools often use controlled electrochemistry at the operating electrode for the analyte's electrochemical reaction [25]. Given the drawback of this passive sensor module, amperometric devices are still believed to achieve a greater sensitivity than potentiometric devices [26].

1.3.4 Conductimetry

This category of sensors focusses on differences in the electrical conductivity of a film or bulk material, the conductivity of which is influenced by the present analyte. Methods of conductimetry are essentially non-selective, hence modification of surfaces and superior equipment has made it more applicable for designing sensors with selectivity. Due to the absence of reference electrodes, conductimetric methods are favourable with good price and easy operating features. One of the ways where a conductimetric method can be used is for calculating the volume fraction of erythrocytes (acting as insulators) present in whole blood. This concept was first employed in the 1940s, and still today it is used in intensive care for measuring haematocrit on multianalyte equipment [27]. Recently, an author has investigated a hydrogen sensor based on covalent carbon nanotube (CNT) paper which was cross-linked with palladium nanoparticles anchored on it [28]. The conductivity was studied by measuring the resistance of the paper in the presence of hydrogen gas. Another researcher used electrospun nylon 6, 6 nanofibre that contained 1-butyl-3-methylimidazolium hexafluorophosphate to obtain a gas sensor acting as a chemiresistor that could detect organic vapours depending on its conductivity [29].

1.3.5 Impedimetry

Sensors based on Impedimetry involve the supply of a voltage that is sinusoidal in nature to measure the current produced rather than a sinusoidal voltage. Impedance is therefore measured in the frequency domain in terms of voltage-to-current ratio [30]. Rheaume and Pisano reported an outstanding study using impedance for gas sensor application [31]. Impedance/capacitance was also used to trace the complex antibody/antigen forming that has been implemented in electrochemical immunosensors. Latest reports of immunosensors driven by impedance involve silicon nitride (Si_3N_4) layer containing anti-human serum albumin (anti-HSA) linked covalently [32]. The electrode structure and function, particularly those altered with biological material, are characterized by a method known as electrochemical impedance spectroscopy (EIS) [33]. When the electrode surface is immobilized, the resistance and capacitance of the double layer fluctuate, triggering an impedance shift. Thus, the mechanism of biorecognition and label-free reactions can be sensed on the sensor surface [34]. A self-assembled monolayer (SAM) or a conductive core layer method based on polymer is usually used to design EIS sensors [35]. It has been reported that reduced graphene oxide shows a very good sensitivity based on impedimetric immunosensor [36].

1.3.6 Field effect transistors (FETs) approach

Field effect transistor (FET) is a class of transistor that requires an electrical field to regulate the channel conductivity between source and drain that act as electrodes in a semiconductor. Conductivity modulation is accomplished by changing the electric field potential at the gate electrode. Under linear mode, when drain-to-source voltage is much less than the gate-to-source voltage, FET works as a switch to swap between conductive and non-conductive states just like a variable resistor provided gate-to-source voltage exceeds drain-to-source voltage. On the other hand, FET can work as a voltage amplifier under saturation mode as it sources out current constantly where the gate-to-source voltage controls it. FET can be converted into a sensor by replacing the gate electrode made of metal with a surface made sensitive to biochemicals (membrane selective to specific analyte or an ion-conductive solution) exposed to analyte solution [37]. A reference electrode exists in the solution, and this leads to the completion of circuit with the help of gate-voltage bias [38].

Over the last few years, the FET structure has been largely acknowledged in real-time biosensing applications as it can detect proteins and nucleic acids by employing its respective receptors as antibodies and nucleic acid conjugates [39–42]. Particular interactivity between analyte molecules and a receptor can detect analytes on a semiconductor or on a neighbouring dielectric surface. Some of the attributes required by a semiconductor for use in biochemically designed sensor applications are high mobility, non-toxins, good stability, longer life cycle, and low-cost production. Both organic and inorganic semiconductors can be used. Examples of organic and/or polymer semiconductors (OSCs) are pentacene, sexithiophene (a-6T), poly(3-hexylthiophene) (P3HT), naphthalene tetracarboxylic diamides (NTCDIs), CNTs, and graphene to name a few whereas in the case of inorganic semiconductors, Si, MoS_2, and SnO_2 are used [43]. Inorganic semiconductors possess higher mobility, better chemical stability, and long-lasting operating time, on the contrary, OSCs provide benefits such as processing latitude and detailed description of the chemical.

Another common method of developing biosensing systems based on FET is to immobilize enzymes on the gate surface of ISFET devices which is pH-sensitive and thus produces an EnFET. The drain current produced is affected by bio-catalytic reactions due to charged particles building up at the surface of the gate. Most popular EnFET implementations include investigations on penicillin, glucose, and urea [26]. Although most of the

ISFET-based sensors were initially introduced for pH testing, most of the FET-based biosensors encounter issues such as reliance of the resulting signal on enzyme immobilization process, characteristics of time response, the complexity of developing effective enzyme-layer deposition methods, and surface patterning techniques consistent with the fabrication of silicon-based instruments [44]. Several more models were proposed for EnFET biosensors [45, 46], as well as for other types of FET-based biosensors, namely FET based on cell, ImmunoFET, GenFET, and so on, commonly classified as BioFET devices or generally BioFETs [44].

FETs can also be considered as an innovative method for the identification and characterization of electrical DNA which proves appropriate for microbial screening and genetic-based disease diagnostics [47]. The variation in DNA material, whether linked to hybridization or enzymatic activity, results in a variation in pH/charge which is localized and a reconfiguration in free ions close to the film surface which regulates the response of the sensor. The far more widely used solutions to analysing DNA hence are based on enzyme-based FED and DNA-modified FED [48].

The latest and the most interesting development in biosensors based on FETs is the use of objects at nanoscale level merged with FET design, including CNTs and nanowires. It consists of a nano-object that connects the source and drain of a FET unit in these kinds of sensing models. Nanowires belong to the category of one-dimensional structures which exhibit attractive properties such as high surface-to-volume ratio, high aspect ratio, high conductivity, and so on. As such, nanowires display very promising bioelectrochemical transducer attributes like relative diameter size with respect to their analysed biochemical analytes wherein their conductivity shows sensitivity towards surface distortion. A pictorial representation of a FET based on silicon nanowire is shown in Figure 1.3 [49]. The electrical performance of the nanowire is altered when particles get attached to it, like voltage and conductivity modulation in conventional FETs. Other

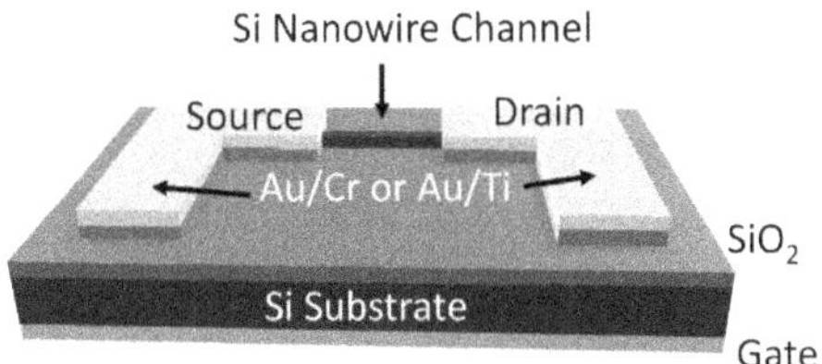

FIGURE 1.3 Illustration of an Si nanowire-based FET.

reports on low dimensional nanostructures for biosensing applications, for example analysis of pH, protein, and DNA binding, microbial and tumour markers integrated into FET systems, have been provided [50–54]. Proper deployment of nanowire-based biosensing devices is contingent on tackling the problems of accurate and replicable positioning during fabrication on mass scale, their functions based on selective biological elements, and their interrelation with biological and electronic systems at nano or micro level.

SUMMARY

The IoT has allowed systems across the globe to store and process information so that they can access it in the future. Nevertheless, due to the relatively long gap in collecting data and the ability in processing and analysing it, this area is still not explored. We explained the vital role that sensors play in IoT to aid in discovering what possibilities ought to be addressed to make the data more meaningful. Over the last few years, apart from promoting IoT communication, the variety of tests that can be carried out using the POCT devices has grown substantially. Reliable and efficient POC diagnoses are capable of revolutionizing public health care. Most of the POCT devices used today use electrochemical sensors. Hence, some of the pre-existing techniques in electrochemistry such as potentiometry, amperometry, conductimetry, and impedimetry used to design and develop POCT devices have been reviewed in depth in this chapter. The advancements and findings in (bio)sensors connecting the biological and engineering sciences are perhaps one of the most interesting scientific fields today. Semiconductor-based technology is evolving so much that we are now seeing a swift influx of advanced sensors based on the latest nanotechnology innovations. We highlighted the applicability of biosensing FET technology with a specific emphasis on the progress being made in integrating existing FET technology and nanotechnology, using structures like nanowires and CNTs. Biosensor development is currently not just powering the fast race to build compact, speedy, affordable, and even more effective sensors but could potentially lead to the full implementation of electronic and biological systems together.

REFERENCES

1. Vermesan, O., P. Friess, P. Guillemin, S. Gusmeroli, H. Sundmaekar, A. Bassi, I. S. Jubert, M. Mazura, M. Harrison, M. Eisenhauer, and P. Doody.

Internet of Things Strategic Research Roadmap in Internet of Things – Global Technological and Societal Trends from Smart Environments and Spaces to Green ICT. Denmark: River Publishers, 2011.

2. López, I. P. *ITU Internet Reports 2005: The Internet of Things.* Geneva, 2005.
3. Lane, N. D., E. Miluzzo, H. Lu, D. Peebles, T. Choudhury, and A. T. Campbell. *IEEE Commun. Mag.* 48 (2010): 140.
4. Sethi, P., and S. R. Sarangi. “Internet of Things: Architectures, Protocols, and Applications.” *J. Elect. Comput. E.* 2017 (2017): 9324035. https://doi.org/10.1155/2017/9324035
5. Sheikh, N. J., and O. Sheikh. In *Proceedings of PICMET ‘16: Technology Management for Social Innovation,* 3082. 2016.
6. Halpern, N. A. *Crit. Care. Clin.* 16 (2000): 623.
7. Brooks, D. E., D. V. Devine, P. C. Harris, J. E. Harris, M. E. Miller, A. D. Olal, L. J. Spiller, and Z. C. Xie. *Clin. Chem.* 45 (1999): 1676.
8. Gubala, V., L. F. Harris, A. J. Ricco, M. X. Tan, and D. E. Williams. *Anal. Chem.* 84 (2012): 487.
9. Song, Y., Y. Y. Huang, X. Liu, X. Zhang, M. Ferrari, and L. Qin. *Trends Biotechnol.* 32 (2014): 132.
10. Severinghaus, J. W. “The History of Blood Gases, Acids and Bases.” *J. Hist. Biol.* 20 (1987): 281.
11. Chaulya, S., and G. M. Prasad. *Sensing and Monitoring Technologies for Mines and Hazardous Areas: Monitoring and Prediction Technologies.* Amsterdam: Elsevier, 2016.
12. D’Orazio, P. *Point of Care.* 3 (2004): 49.
13. Bobacka, J., A. Ivaska, and A. Lewenstam. *Chem. Rev.* 108 (2008): 329.
14. Stradiotto, N. R., H. Yamanaka, and M. V. B. Zanoni. *J. Braz. Chem. Soc.* 14 (2003): 159.
15. Gadzekpo, P. Y., James M. Hungerford, Azza M. Kadry, Yehia A. Ibrahim, and Gary D. Christian. *Anal. Chem.* 57 (1985): 493.
16. Metzger, E., R. Dohner, W. Simon, D. J. Vonderschmitt, and K. Gautschi. *Anal. Chem.* 59 (1987): 1600.
17. Eisenman, G. *Glass Electrodes for Hydrogen and Other Cations: Principles and Practice.* New York: M. Dekker, 1967.
18. Wang, J. *Analytical Electrochemistry.* New York: VCH Publishers, 1994.
19. Pioda, L. A. R., V. Stankova, and W. Simon. *Anal. Lett.* 2 (1969): 665.
20. Freiser, H., ed. *Ion Selective Electrodes in Analytical Chemistry.* Boston: Springer, 1980.
21. Freiser, H. *J. Chem. Soc., Faraday Trans 1.* 82 (1986): 1217.
22. Cunningham, L., and H. Freiser. *Anal. Chem.* 54 (1982): 1224.
23. Power, A. C., and A. Morrin. *Electroanalytical Sensor Technology in Electrochemistry.* Rijeka: InTech, 2013.

24. Chaubey, A., and B. D. Malhotra. *Biosens. Bioelectron.* 17 (2002): 441.
25. Luppa, P. B., L. J. Sokoll, and D. W. Chan. *Clin. Chim. Acta.* 314 (2001): 1.
26. Grieshaber, D., R. MacKenzie, J. Vörös, and E. Reimhult. *Sensors-Basel.* 8 (2008): 1400.
27. Rosenthal, R. L., and C. W. Tobias. *J. Lab. Clin. Med.* 33 (1948): 1110.
28. Ventura, D. N., S. Li, C. A. Baker, C. J. Breshike, A. L. Spann, G. F. Strouse, H. W. Kroto, and S. F. A. Acquah. *Carbon* 50 (2012): 2672.
29. Kang, E., M. Kim, J. S. Oh, D. W. Park, and S. E. Shim. *Macromol. Res.* 20 (2012): 372.
30. Dziąbowska, K., E. Czaczyk, and D. Nidzworski. "Application of Electrochemical Methods in Biosensing Technologies." In *Biosensing Technologies for the Detection of Pathogens–A Prospective Way for Rapid Analysis,* edited by T. Rinken and K. Kivirand, London: IntechOpen, 2017.
31. Rheaume, J. M., and A. P. Pisano. *Ionics* 17 (2011): 99.
32. Caballero, D., E. Martinez, J. Bausells, A. Errachid, and J. Samitier. *Anal. Chim. Acta.* 720 (2012): 43.
33. Yang, L., and R. Bashir. *Biotechnol. Adv.* 26 (2008): 135.
34. Bettazzi, F., G. Marrazza, M. Minunni, I. Palchetti, and S. Scarano. *Compr. Anal. Chem.* 77 (2017): 1.
35. Caygill, R. L., G. E. Blair, and P. A. Millner. *Anal. Chim. Acta.* 681 (2010): 8.
36. Li, X., Y. Wang, X. Zhang, Y. Gao, C. Sun, Y. Ding, F. Feng, W. Jin, and G. Yang. *Microchim. Acta.* 187 (2020): 217.
37. Errachid, A., N. Zine, J. Samitier, and J. Bausells. *Electroanal.* 16 (2004): 1843.
38. Schöning, M. J., and A. Poghossian. *Analyst* 127 (2002): 1137.
39. Chen, K., B. Li, and Y. Chen. *Nano Today* 6 (2011): 131.
40. Joshi, R. K., and J. J. Schneider. *Chem. Soc. Rev.* 41 (2012): 5285.
41. Sarkar, T., Y. Gao, and A. Mulchandani. *Appl. Biochem. Biotechnol.* 170 (2013): 1011.
42. Zhou, Y., C. W. Chiu, and H. Liang. *Sensors* 12 (2012): 15036.
43. Huang, W., A. K. Diallo, J. L. Dailey, K. Besar, and H. E. Katz. *J. Mater. Chem. C* 3 (2015): 6445.
44. Schöning, M. J., and A. Poghossian. *Electroanal.* 18 (2006): 1893.
45. Yin, L. T., J. C. Chou, W. Y. Chung, T. P. Sun, and K. P. Hsiung. *Sensor Actuat B-Chem.* 76 (2001): 187.
46. Poghossian, A., A. Cherstvy, S. Ingebrandt, A. Offenhausser, and M. J. Schoning. *Sensor Actuat B-Chem.* 111 (2005): 470.
47. Ghosh, I., C. I. Stains, A. T. Ooi, and D. J. Segal. *Mol. BioSyst.* 2 (2006): 551.
48. Veigas, B., E. Fortunato, and P. V. Baptista. *Sensors* 15 (2015): 10380.
49. Koo, S. M., M. D. Edelstein, Q. L. Li, C. A. Richter, and E. M. Vogel. *Nanotechnology* 16 (2005): 1482.

50. Hakim, M. M. A., M. Lombardini, K. Sun, F. Giustiniano, P. L. Roach, D. E. Davies, P. H. Howarth, M. R. R. De Planque, H. Morgan, and P. Ashburn. *Nano Lett.* 12 (2012): 1868.
51. Cheng, Y., K. S. Chen, N. L. Meyer, J. Yuan, L. S. Hirst, and P. B. Chase. *Biosens. Bioelectron.* 26 (2011): 4538.
52. Park, S. J., O. S. Kwon, S. H. Lee, H. S. Song, and T. H. Park. *Nano Lett.* 12 (2012): 5082.
53. Kuang, Z., S. N. Kim, W. J. Crookes-goodson, and B. L. Farmer. *ACS Nano.* 4 (2010): 452.
54. Lerner, M. B., J. D'Souza, T. Pazina, J. Dailey, B. R. Goldsmith, M. K. Robinson, and A. T. C. Johnson. *ACS Nano.* 6 (2012): 5143.

CHAPTER 2

Preparation of Metal Oxide Semiconductor Nanostructure and Their Structural Properties

Paulsamy Chinnamuthu, Iman Biswas, and Aniruddha Mondal

2.1 INTRODUCTION

In the last decade, an accumulative growth of research activities has been seen in nanoscience and nanotechnology [1–6]. Depending on the carrier confinement, nanostructures can be categorized as (1) Quantum dots (0-D), (2) Quantum wires (1-D), and (3) Quantum wells (2-D). Figure 2.1 illustrates the schematic representation of an electron system in bulk structure and low-dimensional nanostructures.

In quantum dots, the carriers are confined in all three dimensions, and in wire (NW) the carriers are allowed to travel only along the axis of the wire, where the other two directions are confined. In the quantum plane, the carriers are allowed to move in any two directions, where the other dimension is confined. The carriers are confined within a thin layer in the case of a quantum plane. The confinement in a particular material system will

DOI: 10.1201/9781003464211-2

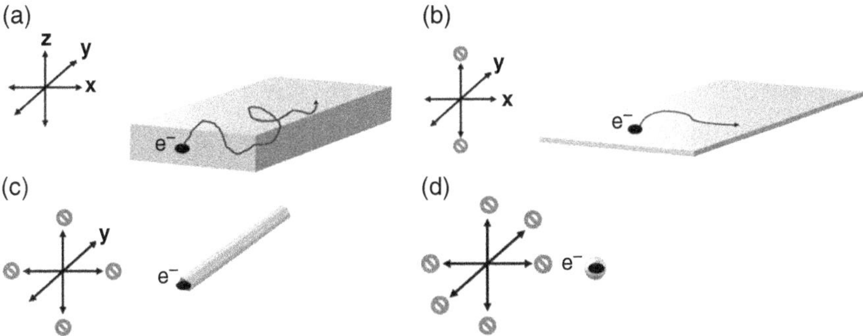

FIGURE 2.1 Schematic illustration of electron system in (a) bulk structure, (b) quantum plane, (c) quantum wire, and (d) quantum dot.

be achieved only when any of its dimensions are equivalent to or smaller than the de Broglie wavelength (≤20 nm) of the carriers. Among the three, the NWs are attracted for their efficient transport of electrons in a particular direction and low exciton interaction in the transport mechanism. The unique physical and chemical properties come out when the size of the material becomes smaller and smaller, and down to the nanometre scale. In a very impressive manner, the specific surface area and surface-to-volume ratio increase as the size of a material decreases. This unique property opens new opportunities in nanoscale electronics and optoelectronics device applications. The quantum confinement in NWs produced phonon confinement, reduced unnecessary phonon-assisted non-radiative recombination process, and introduced high carrier mobility. The small active area of the NW devices can yield higher light sensitivity and fast switching of the detector. One of the most studied phenomena in NWs is their sensitivity to light (photoconductivity).

As an important type of optoelectronic device, ultraviolet (UV) photodetectors show wide applications in space-to-space communications, environmental sensors, water sterilization, flame sensing, early missile plume detection fire monitoring, and UV irradiation detection [7–11]. All these applications require very sensitive devices with a high signal-to-noise ratio and high response speed. A variety of UV devices are available, mainly Si-based photodetectors and photomultipliers. These devices can be very sensitive in the UV region with low noise and quick response. However, they have significant limitations, such as the need for filters to stop low energy photons (visible and IR light), their degradation and lower efficiency (Si-based photodetectors), or the need for an ultra-high vacuum environment and a very high voltage supply (photomultipliers) [12]. To address this issue, UV detectors based on wide bandgap semiconductors such as diamond, SiC, GaN, and ZnO materials have drawn much attention due to their intrinsic visible blindness. Moreover, wide-bandgap materials are chemically and thermally more stable, which is an advantage for devices operating in harsh environments [13]. Among these materials, III-nitrides, SiC, and ZnO are the most widely studied semiconductors [14–18], but their domestic utilization is hampered by their high cost and complicated fabrication technologies [19, 20]. Recently, TiO_2-based UV photodetectors have attracted considerable interest. TiO_2 is a wide-band semiconductor, which has very good physical, chemical, and optical properties. The bandgap of TiO_2 is 3.0 eV ~ 3.2 eV, which makes it very attractive for the UV detector. In recent years, significant progress has been made in improving the performance of TiO_2-based UV photodetectors. However, some serious problems remain due to its low operating frequency and unable to compete with Si and III-N semiconductor-based devices. However, TiO_2 material is attractive for its non-toxicity [21], chemical stability [22], and low-cost fabrication procedure [23]. The problems defined with the TiO_2 devices can be abolished if someone can grow a crystalline TiO_2 material having infinitesimal oxygen defects.

Also, the ability to form heterostructures with modulated composition and/or doping enables the passivation of interfaces and the generation of devices with diverse functions. These interfaces include dielectric–semiconductor junctions, semiconductor heterojunctions, p-n homojunctions, and metal-semiconductor junctions. A heterojunction of two materials can be an undoped semiconductor to an undoped or a doped semiconductor or a junction between regions having different dopant concentrations. The recent advances in their design and control techniques in the formation of heterostructures have led to some new device concepts. Moreover, 1-D semiconductor heterostructures consisting of two or more components with distinct functionalities can exhibit new or enhanced performances, such as emission efficiency and high electron mobility, which are the key factors for many devices. In particular, the major advantage of heterostructures is that different materials have different bandgaps which make it very important for potential future applications in modern electronics and optoelectronics applications. Thus, the introduction of complex heterostructures into semiconductor NWs is a necessary step for their efficient exploitation in numerous applications.

Before starting the device fabrication procedure and its characterization, in the next session, we have described the different synthesis procedures, which will help the reader to understand the reality of simple and low-cost methods for further modification to solve the actual problems in device operation.

2.2 SYNTHESIS METHODS OF METAL OXIDE NANOWIRES

The unique properties of nanostructured materials have been described in the above section which has led many researchers to pursue synthesis, doping, and novel device applications of metal oxide semiconductors due to their potential applications in nanoscale electronic and optoelectronic devices [24–28]. Various research groups have developed different methods to fabricate the nanostructure using electron beam lithography, plasma etching, molecular beam epitaxy (MBE), metal-organic chemical vapour deposition (MOCVD), plasma-enhanced chemical vapour deposition (PECVD), chemical vapour deposition (CVD), vapour-liquid-solid (VLS), physical vapour deposition (PVD) techniques, and some chemical synthesis methods. In general, fabrication techniques for nanostructures can be broadly divided into two categories: top-down and bottom-up.

2.2.1 Top-Down Methods

Top-down approaches begin with bulk material from which nanostructures are patterned via a combination of lithography and etching, for example, using electron beam lithography and plasma etching or focussed ion beam milling. Top-down methods have underpinned the microelectronics industry to date, but as the length scales of devices shrink according to Moore's law, top-down methods become increasingly problematic.

The lithographic and etching techniques make it difficult to define smaller features, and the quality of the nanostructures diminishes. The etching and patterning processes introduce surface defects, which adversely affect nanostructure properties. Furthermore, the process is intrinsically wasteful, and there are large challenges to overcome before these technologies become high-throughput, cost-effective means of NW fabrication. The schematic representation of the top-down process is shown in Figure 2.2.

2.2.2 Bottom-Up Methods

The bottom-up approach refers to the build-up of material from the bottom: atom-by-atom, molecule-by-molecule, or cluster-by-cluster. This

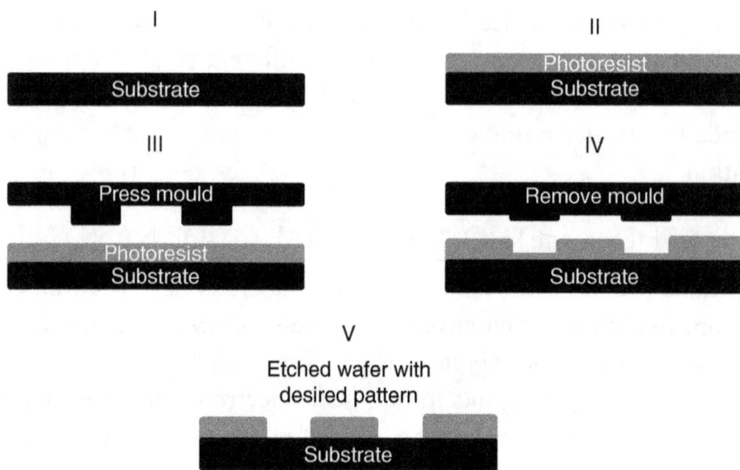

FIGURE 2.2 Schematic representation of top-down process.

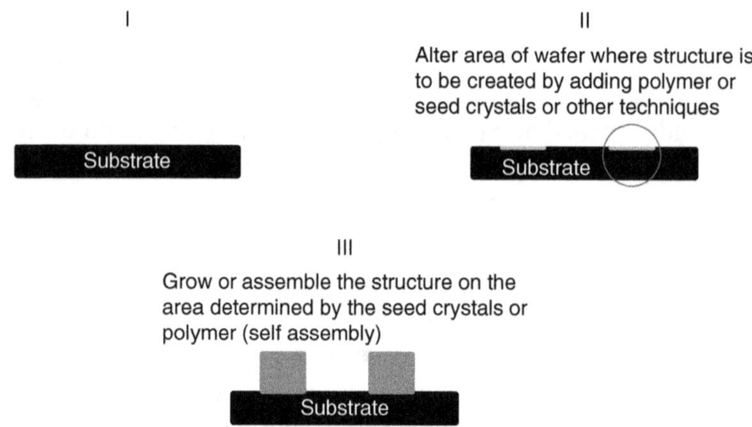

FIGURE 2.3 Schematic representation of bottom-up process.

bottom-up paradigm offers opportunities for the fabrication of atomically precise, complex devices that are not possible with conventional top-down technologies. The most common, as well as the most promising, techniques used for the fabrication of nanostructures are MBE, CVD, and PVD. Figure 2.3 shows the general process diagram of the bottom-up design.

2.2.2.1 Physical Vapour Deposition

Vapour deposition refers to any process in which materials in a vapour state are condensed to form a solid-phase material. PVD is a simple technique, and it is suitable for the large area-controlled growth [29].

2.2.2.2 *Oblique Angle Deposition and Glancing Angle Deposition Techniques*

The available reports on one-dimensional (1D) nanostructure growths are restricted by template growth technique [30, 31]. Gold is the most used material for particle-assisted 1D nanostructure growth. It is a challenge to grow the symmetric NWs. MBE [32], MOCVD, [33] PECVD [34], and CVD [35] are popular techniques to grow the III-V 1D nanostructure, which are not cost-effective. VLS [36] and some chemical synthesis methods are employed for 1D oxide nanostructure growth. The VLS is a catalytic employed growth technique. The catalysts like Ag, Au, Pt, and so on helped to grow the NW and have disadvantages to separate them at the time of device fabrication. Recently, several nano-lithographic [30] techniques for the fabrication of controlled 1D structures are available. However, the techniques are not cost-effective for large-area 1D nanostructure growth. Whereas the unconventional methods based on chemical synthesis are cost-effective and have the potential for large volume growth [37]. It is very difficult to control the morphology of the nanostructure in the chemical synthesis method. The oblique angle deposition (OAD) and glancing angle deposition (GLAD) are comparatively low cost than nanolithography and are more suitable for large area-controlled growth [29]. The techniques are one-step catalytic free growth techniques and can be developed inside the chamber of a PVD machine like thermal evaporator, electron-beam evaporator, and sputtering.

Developments in OAD technology during the last three decades have produced columnar nanostructures of various shapes (including vertical, tilted, helical, and chevron) and graded-porosity thin films for use in applications ranging from sensors and actuators to optical filters, microfluidics, and catalysis. The growth of the 1D structure in OAD is dominated by vapour flux angle [38]. The conditions of limited adatom diffusion in OAD enhance atomic shadowing and create an inclined 1D structure. 1D nanostructure of different shapes has great attention for fabricating electronics, optoelectronics, electrochemical, and electromechanical devices [39]. The vapour mean free path should be larger than the source-substrate distance to fabricate high-quality structures. This is easily achieved with electron-beam evaporation systems, widely used for GLAD depositions.

OAD and GLAD are bottom-up processes that enable the fabrication of nano-engineered films. The vapour created is largely collimated so that the grown thin films (TF) have a columnar morphology. In the past

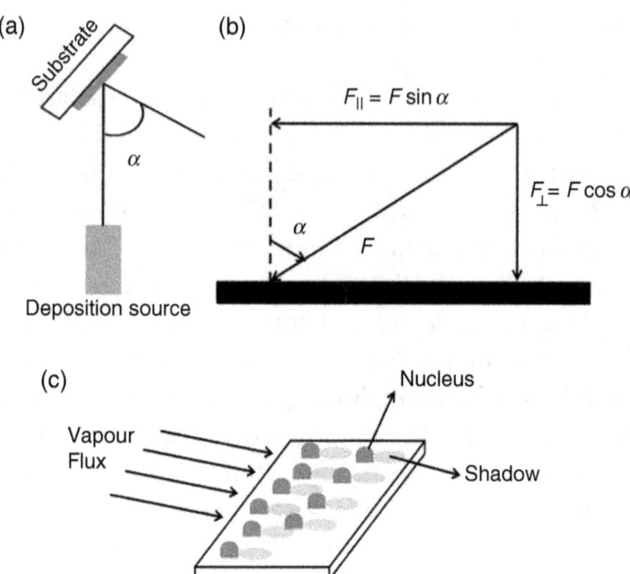

FIGURE 2.4 (a) Experimental arrangement for OAD (schematic), (b) evaporated incident flux (F) material on the substrate and its components and (c) shadowing effect.

ten decades, the OAD has been of interest to the TF community for the enhancement of properties such as dichroism, birefringence, and anisotropic resistivity [40–43].

The film microstructure which produces these novel properties and capabilities depends on ballistic shadowing and the formation of columnar microstructures during deposition. The experimental setup for OAD is very simple, which is shown in Figure 2.4(a). If we can treat the incoming vapour flux as a vector F as shown in Figure 2.4(b), the flux has two components: a vertical component $F_{\perp} = F \cos \alpha$ and a lateral component (a vector) $F_{||} = F \sin \alpha$. The substrate will receive the vapour flux from both the vertical and lateral directions.

During the deposition of TF onto a flat substrate, initially, the impinging atoms [Figure 2.4(c)] will randomly form islands on the substrate. As deposition proceeds, the initial nucleated islands will act as shadowing centres, and all the tallest islands will receive more impinging atoms as compared to the shorter ones (shadowing effect). This competition process will only leave the tallest islands to grow into columns, and a nanocolumnar film will be formed. The lateral component $F_{||}$ is the source of the shadowing effect.

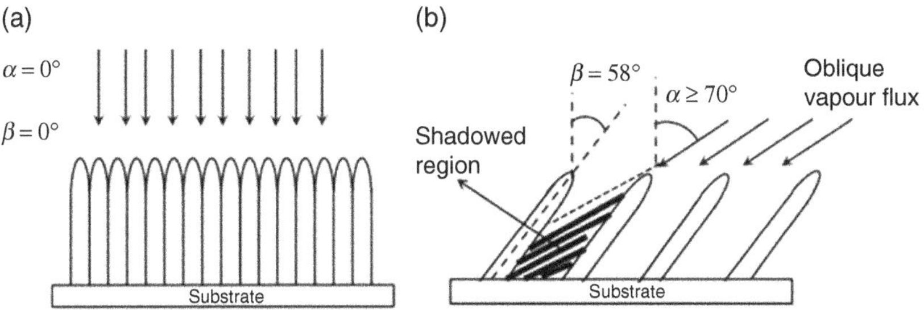

FIGURE 2.5 Schematic diagram of nanostructure by OAD method: (a) Very low porosity (TF); (b) separate column (Slanted).

For the OAD, since $F_{\parallel}$ remains constant during deposition, a columnar film with a tilt angle β will be formed. In general, the column tilt angle β is less than the vapour incident angle α and follows the empirical tangent rule [44], $\tan\beta = 1/2\tan\alpha$ for small $\alpha > 50°$, or the Tait relation [45] $\beta = \alpha - arc\ \sin(1-\cos\alpha)/2$ when the α is large. If the incidence angle of vapour flux $\alpha = 0°$, then the substrate is directly pointed at the source. The schematic diagram is shown in Figure 2.5(a), with very low porosity, which is nothing but TF. Slanted posts with high porosity occur when a substrate is held stationary during the deposition. For isolated posts, α is typically $\geq 70°$ [Figure 2.5(b)].

GLAD [Figure 2.6(a)] is an extension of OAD where the substrate position is manipulated during film deposition. Although some early experiments used mobile substrates [46], oblique depositions are typically performed with a fixed substrate. GLAD developed as researchers realized they could manipulate the columnar structure by actively managing substrate position during deposition.

Ballistic shadowing is the foundation of GLAD engineering. Shadowing is possible if only the incoming vapour flux is well collimated. If there is a large angular spread in incoming vapour flux, shadows will be poorly defined. Two main approaches to achieve collimated vapour flux are: the large distance between vapour source and substrate, and physical obstacles that select a subset of an uncollimated vapour plume. For a fixed substrate size, as the distance from the source increases, incident vapour collimation improves. However, the number of collisions also increases the farther the vapour flux travels before reaching the substrate.

A flexible GLAD system should have a mean free path longer than the chamber dimensions. For typical systems with a source-substrate distance

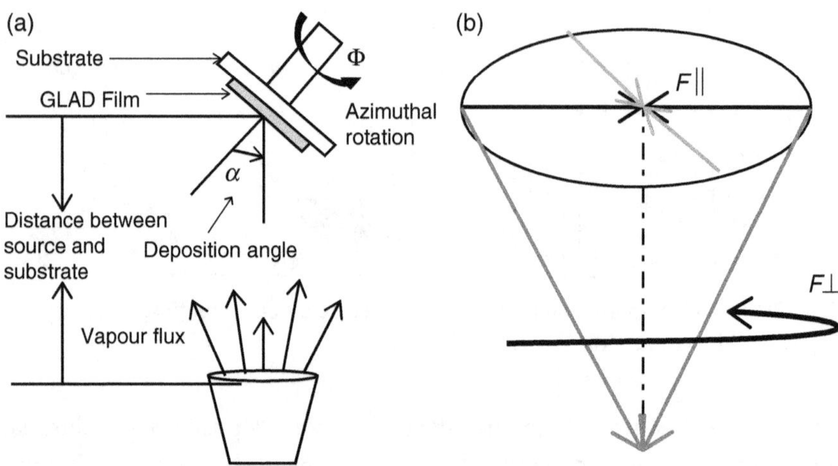

FIGURE 2.6 Schematic diagram of (a) GLAD with incident angle α of the vapour flux and substrate azimuthal (ϕ) rotation and (b) cancelling of $F_{\|}$ components in GLAD.

of ~ 45 cm, an operating pressure of $\leq 10^{-3}$ torr is required. While these constraints vary with the source type used, in general, a lower operating pressure is preferred to produce high-quality structures. Many vapour sources have been used for GLAD, including electron-beam evaporation, thermal evaporation, sputtering, several co-sputtering techniques, and pulsed-laser deposition.

The azimuthal rotation of the substrate with respect to substrate surface normal will be controlled by the motor. During the deposition, the substrate will receive the vapour from both the components, but for the growth of the vertically aligned NWs, the deposition should be preceded by $F_{\perp}$ only, that is, we have to cancel out the $F_{\|}$ component. This forms the basis of the GLAD technique for the growth of vertically aligned NWs. When the substrate rotates azimuthally, each part has an equal chance to receive the same amount of particle from the $F_{\|}$ component. After a complete revolution, the average $\sum F_{\|}$ is zero due to the cancellation of $F_{\|}$ components [Figure 2.6(b)] at the opposite directions which means that there is no preferred orientation of the NWs. By this technique, purely vertical NW will grow on the substrate surface.

The deposition process in GLAD is complicated, and the column structure depends on the angle of vapour incidence (α), substrate rotation (ϕ), type of rotation and direction of rotation, and even on the rate of deposition. The behaviour of both α and ϕ is broken into three categories: constant, discrete, and continuous. A constant designation means that the substrate remains stationary at a given angle during a deposition, that is, OAD. A discrete designation means that the substrate undergoes periodic changes in a given angle but remains stationary otherwise. A continuous designation means the substrate is in motion continuously during a deposition for the specified angle. In the forthcoming session, we will discuss the GLAD technique used for the synthesis of metal oxide NWs.

2.2.3 Synthesis of Metal Oxide NWs Using GLAD

The GLAD technique has been employed for the synthesis of Metal Oxide NRs. The WO_3 were deposited by the DC magnetron sputtering at a base pressure of ~7.0×10^{-6} mbar, the operating pressure of 5.6×10^{-3} mbar with O_2 (99.999%) flow rate of 20 sccm. [47] Both the growth rate and film thickness were monitored by quartz crystal inside the chamber. The distance between the target and the substrate was maintained at 7 cm.

The FESEM images shown in Figure 2.7(a) to (g) illustrate that adjusting the rotation speed and deposition angle allows for the precise formation

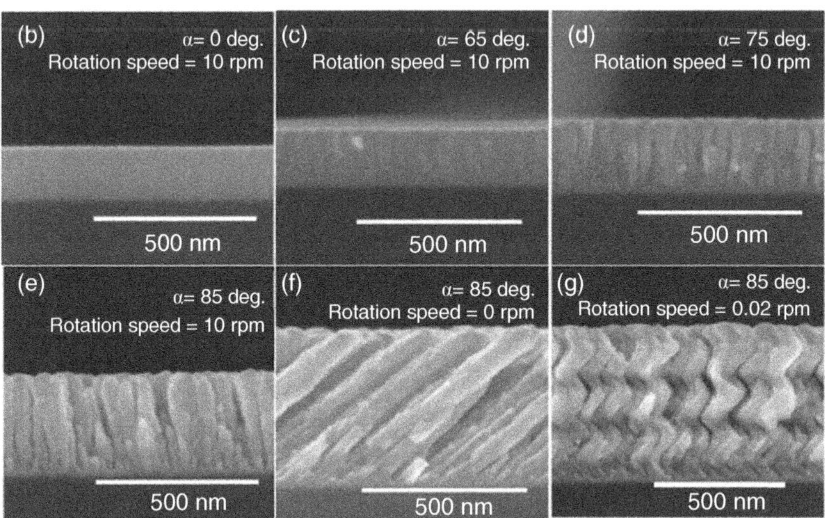

FIGURE 2.7 FESEM images of WO_3 with different deposition angles and rotation speeds. [Reproduced from M. Horprathum, P. Eiamchai, J. Kaewkhao, C. Chananonnawathorn, V. Patthanasettakul, S. Limwichean, N. Nuntawong, P. Chindaudom; Fabrication of nanostructure by physical vapour deposition with glancing angle deposition technique and its applications. AIP Conf. Proc. 25 September 2014; 1617 (1): 7–11. https://doi.org/10.1063/1.4897091; Figure 3 with the permission of AIP Publishing.]

of vertically aligned nanorods. The substrates were used at a constant azimuthal rotation of 0 rpm, 10 rpm, and 0.02 rpm and orientation of 0°, 65°, 75°, and 85° with respect to the perpendicular line between the material source and the planer substrate holder. The comparison between α of 75° and 85° reveals that altering the angle impacts the porosity, with the 85° angle exhibiting higher porosity compared to the 75° angle [Figure 2.7(d) and 2.7(e)].

Nanostructures with a spiral shape were observed when the substrate rotated at an extremely slow rate of 0.02 rpm, as depicted in Figure 2.7(g), due to the continuous variation in the direction of the incoming vapour flux, facilitated by the slow rotation speed. In contrast, without substrate rotation, the shadowing effect occurred in only one direction, causing the columns to slant towards the source [Figure 2.7(f)].

The top view of the sample was prepared at 85° GLAD as shown in Figure 2.8(b). A clear view of the NW top has been observed with the FE-SEM image. In the case of 70° GLAD, the low atomic shadow was not able to form the NWs on the Si substrate, whereas 85° GLAD created a large shadow and was suitable to create an NW array on the substrate. The side view [Figure 2.8(c)] of the 85° GLAD sample shows the perpendicular columnar structure of TiO_2 on Si. Highly periodic and ordered perpendicular SiO_x-TiO_2 multi-layered NWs with a diameter of 50 nm on ITO-coated glass substrate [48] have been synthesized. Figure 2.8 shows the FEG-SEM images of the as-deposited samples prepared at 85° GLAD and 460 rpm azimuthal rotation of the ITO substrate. The top view [Figure 2.8.(a)] of NWs shows that some of the NWs (marked by dashed circles) have left their growth before attaining their full lengths, which can be attributed to a competitive growth mode procedure during the deposition. The average top diameter of the NWs calculated was ~50 nm. The individual NWs clumped into each other and formed the clusters. Figure 2.8(b) shows the cross-sectional view of the sample, which contains a perpendicular columnar structure of NW arrays with an average height of ~450 nm. The EDAX analysis [Figure 2.8(c)] shows the presence of titanium (Ti), oxygen (O_2), silicon (Si), and indium (In) in the samples. The spectrum shows the emission from Ti K, O_2 K, Si K, and In L shells. The emission from In L shell was from the ITO. Figure 2.8(d) shows the EDX mapping of the FEG-SEM image of the sample as well as the presence of Ti, O_2, Si, and In. In the GLAD technique, some of the SiO_x NWs grew smaller during their deposition as a consequence of the competitive growth mechanism. For further deposition

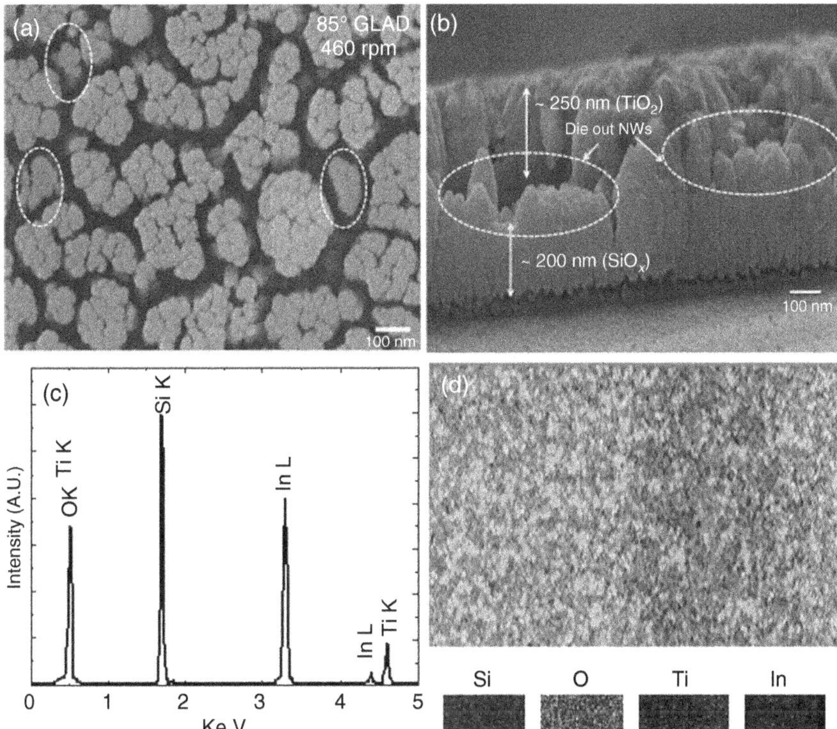

FIGURE 2.8 FEG-SEM image of GLAD SiO_x-TiO_2 heterostructure NW arrays: (a) top view, (b) cross-sectional view, (c) EDAX analysis, (d) chemical mapping [48].

of TiO_2, taller SiO_x NWs captured most of the TiO_2 flux and acquired the SiO_x-TiO_2 double-layered NWs. Hence, in that continuous process of deposition, the smaller sub-NWs eventually die out [Figure 2.8(b)] due to the atomic shadowing of overgrown larger neighbours.

Figure 2.9 shows the schematic representation of the formation of heterostructure NWs together with the sub-grown NWs region. In the context of the above growth mechanism, the average lengths of sub-grown NWs for SiO_x-TiO_2 heterostructure were calculated at ~200 nm from the highlighted portion of the cross-sectional FEG-SEM image [Figure 2.8(b)].

Figure 2.10(a) shows that the NWs are nearly symmetrical at both the top and the bottom having a width of ~50 nm. The bottom of the NWs consists of $SiO_{x,}$ and the top consists of TiO_2, clearly viewed from the difference in colour contrast of a typical TEM image of the NWs [Figure 2.10(a)]. The

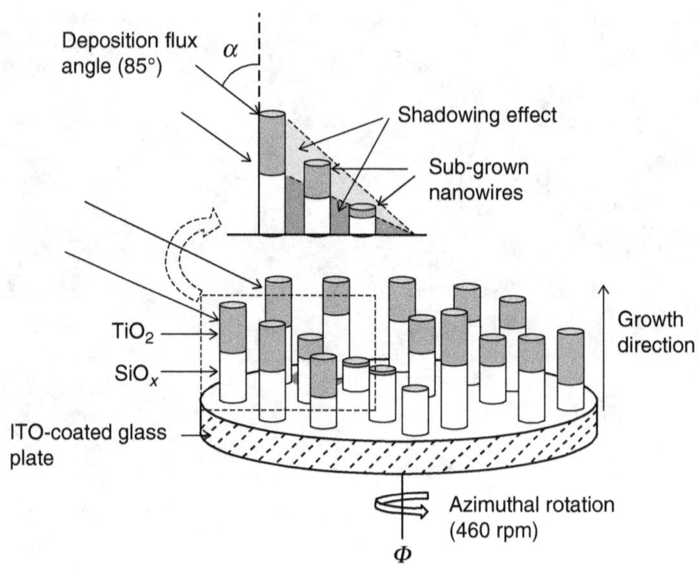

FIGURE 2.9 Schematic diagram of the heterostructure NWs formation [48].

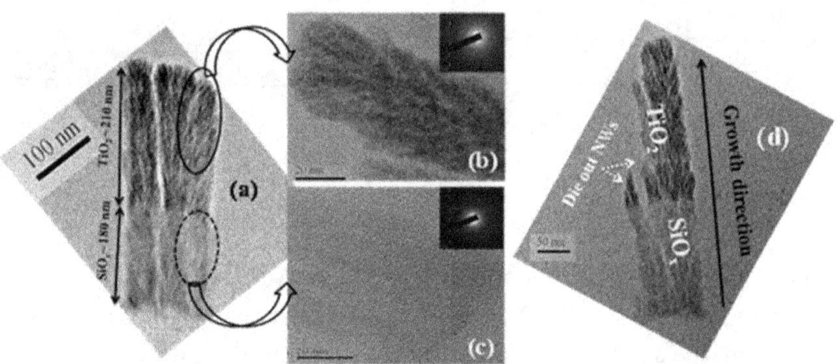

FIGURE 2.10 TEM images of the SiO_x-TiO_2 heterostructure NW: (a) separate NWs; (b) upper portion (TiO_2) of the heterostructure with SAED (inset); (c) lower portion (SiO_x) of the heterostructure with SAED (inset); and (d) typical sub grown NWs [48].

lighter portion of the NWs is SiO_x of length ~180 nm and comparatively the darker portion of the NWs is TiO_2 of length ~210 nm. Figures 2.10(b) and 2.10(c) show that the deposited SiO_x and TiO_2 are both amorphous in nature, which was confirmed by SAED analysis. The growth direction of the NWs is shown in Figure 2.10(d). The NWs consist of tiny TiO_2 on SiO_x due to a competitive growth mechanism, also shown in Figure 2.10(d).

2.2.4 Vapour-Liquid-Solid Deposition for Metal Oxide NWs Synthesis

Using this method, 1D nanostructures like NWs, nanotubes, nanorods, and whiskers can be synthesized. R.S Wagner and W.C Ellis in 1964 explained a crystal growth technique in which a liquid layer is used between the melt and the growing crystal [49].

Here at first, a catalyst is deposited. This catalyst gradually forms liquid droplets at high temperatures. The target material is supplied through the vapour on the liquid. Therefore, the liquid gets supersaturated, and the precipitation of this supersaturated solution at the solid–liquid interface results in crystal growth in one direction. Since this process involves vapour, liquid, and solid phases, it is called a vapour-liquid-solid (VLS) technique of crystal growth. The liquid must be stable, and this gives a minimum diameter of the nanostructure grown. Furthermore, it can be well understood that the diameter of the 1D nanostructure depends upon the diameter of the catalyst nano drops, and the catalyst remains at the top of the 1D nanostructure at the end of the process. To choose the catalyst material for the crystal growth, the phase diagram of the target material is studied and the catalyst composition and the temperature required for the growth are decided. Figure 2.11 explains the VLS growth process where

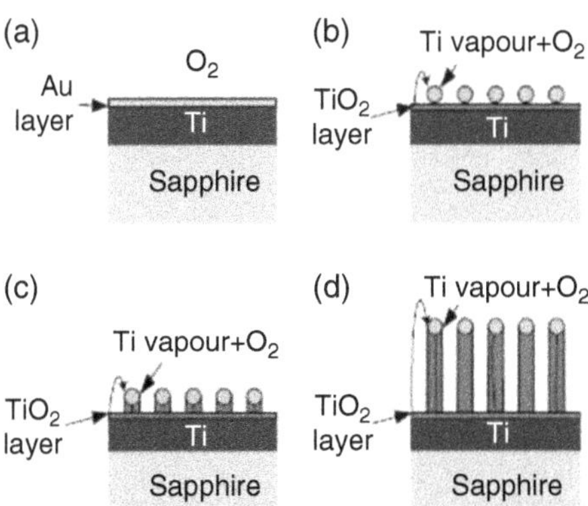

FIGURE 2.11 Schematic illustration of the growth procedure of TiO_2 NWs on a sapphire substrate. (a) The metal-droplet-coated substrate is exposed to reactant precursors; (b) growth of a monophase NW with the metal droplet acting as catalyst; (c) a superlattice NW has been developed; and (d) a coaxial NW finally formed. [Reproduced from Lee, Jung-Chul; Park, Kyung-Soo; Kim, Tae-Geun; Choi, Heon-Jin; Sung, Yun-Mo; Controlled growth of high-quality TiO2 nanowires on sapphire and silica. Nanotechnology. 2006; 17, 4317–4321. Figure 4 with the permission of IOP Publishing].

the metal-droplet-coated substrate is exposed to reactant precursors and a monophase NW is grown with the metal droplet acting as a catalyst.

The semiconductor NW can be converted into a heterostructure NWs by adding one or more layers of another semiconductor horizontally or radially to it by varying the precursor used. This gives either superlattice NW or coaxial NW [50]. In the past decade, researchers have grown TiO_2 NWs using the VLS mechanism. But it is very important to select the appropriate metal catalyst. The catalyst used for the VLS mechanism should fulfil the following requirements as mentioned by Wagner et al. [49].

(1) At the deposition temperature, which is between 800°C and 1150°C for TiO_2 NW growth using the VLS process, a catalyst must be able to form a liquid solution with this material to be grown.

(2) The relation between the solubility of the catalyst in the solid and liquid must be such that its ratio, known as the distribution coefficient, is

    $K = C_s/C_l < 1$, where C_s is the solubility of the catalyst in solid and C_l is the solubility of the catalyst in liquid.

(3) There must be a small vapour pressure of the catalyst over the liquid alloy to prevent the evaporation of the catalyst. The evaporation of the catalyst affects the volume as well as the cross-section of the crystal.

(4) The catalyst should not react to the products of the chemical reaction.

(5) The deposition temperature should be so selected that the catalyst does not form an intermediate solid phase.

(6) The growth of the crystal material from the metal catalyst must be faceted which is possible if the growth occurs from a dilute solution. So, researchers have used different metal catalysts for titanium dioxide NW growth. Jung-Chul Lee et al. reported the growth of TiO_2 NW using the VLS method on both silica and sapphire substrates [51]. The first deposited a titanium layer which served as a buffer layer between the substrate and the Ti vapour source to avoid diffusion of the Ti vapour into the substrate. After that, Au of 50 nm thickness was deposited on it. These Au nanoparticles were used as a catalyst. Ti vapour along with the O_2 reactive gas was supplied on it. The Titanium vapour supplied from the Ti buffer layer as well as the Ti source powder to Au catalyst initiated the growth of TiO_2 NW as shown in Figure 2.11.

The TiO_2 NW grown on the quartz substrate showed high-quality crystallinity than that grown on the sapphire substrate. The authors reported the NWs grown on the sapphire substrate with a diameter ranging from 80 to 100 nm and length of 1–5 μm and on the quartz substrate with a diameter of 50–80 nm and length of 1–3 μm, respectively.

Lin Yue et al. reported the growth of TiO_2 nanotubes [52]. The catalyst used was Ceria (CeO_2) nanoparticles which were synthesized using the hydrothermal method. The initial material used for the CeO_2 was the mixture of Ce $(SO_4)_2$ and ammonia solution. Ti $(SO_4)_2$ (titanium sulphate) was added to the Ceria nanoparticles at 303 K. As the time increased, the length of the nanotube also increased. Rectangular titania NW [53] of diameter 50–500 nm and length 3–8 μm were grown using tin (Sn) as the catalyst and by supplying Ti from the patterned substrate and oxygen from the reactor ambience as shown in Figure 2.12.

Other parameters like the thickness of the metal catalyst layer, the gas flow rate, or the growth temperature when varied showed some changes in the morphology of the TiO_2 NW grown.

A research group fabricated TiO_2 NW for Osseointegration using Sn as a metal catalyst [54]. They deposited Sn catalysts of different thicknesses of 10, 30, and 50 nm and studied the effect of this thickness on the diameter

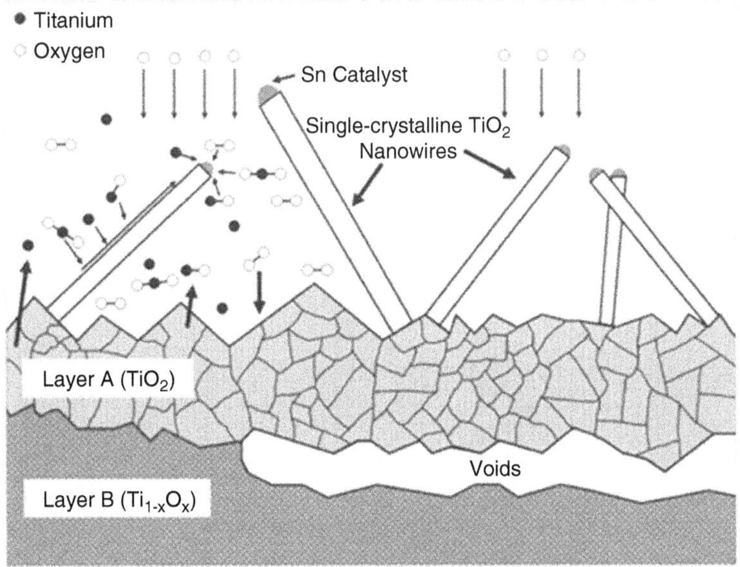

FIGURE 2.12 Growth of TiO_2 NW on Sn catalyst using the VLS mechanism [53].

and structure of the TiO_2NW grown. Figure 2.13 shows the variation of the diameter of TiO_2 NW with the change in thickness of the Sn layer. They reported that the surface of the TiO_2 NW grown on the 10 nm Sn layer was smooth compared to that grown on 50 nm thickness.

The ZnO NWs synthesized by Dongshan Yu used a modified vapour phase transport deposition process on fluorine-doped tin oxide (FTO) substrates [55]. The synthesis process involves a unique double-tube reactor configuration that creates a Zn-rich vapour environment conducive to the formation and growth of zinc oxide NWs. The growth of ZnO NWs is influenced by several key parameters, including reaction time, temperature, and carrier gas flow rate, which have a significant impact on the size, morphology, crystalline structure, and density of the NWs. Longer reaction times lead to an increase in the average diameter and length of ZnO NWs, with nucleation and growth occurring from ZnO particles formed during the process. The carrier gas flow rate significantly impacts the number density of synthesized NWs, with experiments conducted at flow rates of 70, 139, and 200 sccm under fixed conditions revealing varying results. A flow rate around 70 sccm is optimal for synthesizing typical ZnO NWs as shown in Figure 2.14, while higher flow rates, such as 200 sccm, lead to the formation of ZnO NP instead of NWs due to a degraded Zn-rich environment. The number density of NWs decreases with increasing flow rate, as observed when comparing NWs synthesized at 70 sccm to those at 139 sccm, where fewer NWs are produced. Similarly, higher temperatures, such

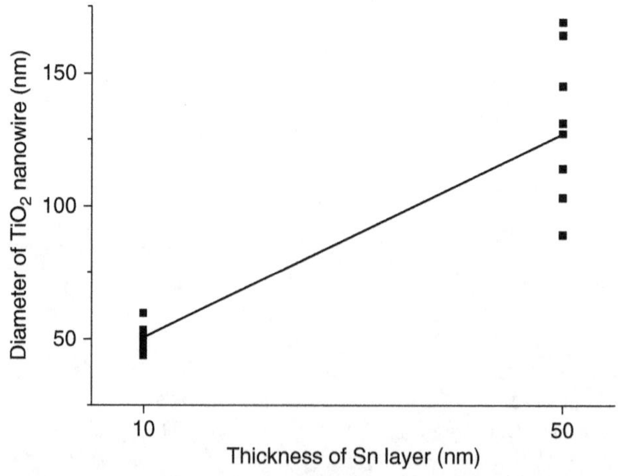

FIGURE 2.13 Variation of the diameter of the TiO_2 NW with Sn catalyst layer thickness [54].

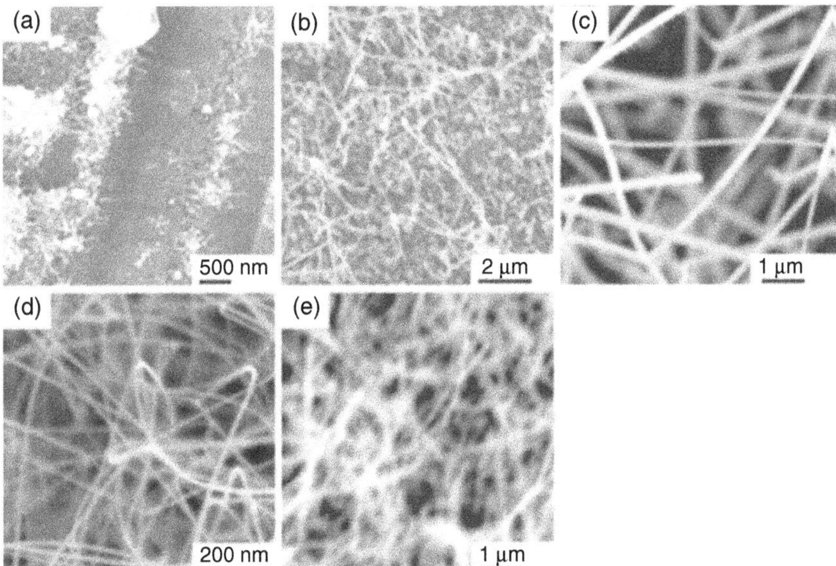

FIGURE 2.14 SEM images of ZnO nanostructures synthesized at temperatures of (a) 910°C, (b) 930°C, (c) 940°C, (d) 950°C, and (e) 1,000°C for 100 minutes, with an Ar flow rate of 70 sccm at 760 Torr [55].

as 950–1000°C, result in thinner and longer NWs, with diameters around 30–50 nm and lengths exceeding 5 μm. Moreover, the size and distribution of nucleation sites during the growth process can impact the diameter and length of the ZnO NWs. Longer reaction times can lead to an increase in the size of nucleation sites, affecting the final characteristics of the NWs. To achieve desired ZnO NW characteristics, a balance of reaction parameters is crucial. Higher temperatures promote longer and thinner NWs with denser growth, while an appropriate carrier gas flow rate ensures optimal nucleation and growth conditions.

2.2.5 Sol-Gel Technique

Sol-gel is a wet chemical-based self-assembly process for nanomaterial formation. The sol-gel process, as the name implies, involves the evaluation of networks through the formation of a colloidal suspension (sol) and gelation of the sol to form a network in a continuous liquid phase (gel). The precursors used for synthesizing the collides generally consist of metallic ions and ligands, which are elements surrounded by various reactive species.

Few alkoxides are immiscible in water; in that situation, some homogenizing agents such as alcohols are used. In general, sol-gel formation occurs in four steps:

(1) Hydrolysis of the precursor in the acidic or basic mediums
(2) Polycondensation of the hydrolysed products
(3) Growth of particles
(4) Agglomeration of particles followed by the formation of networks throughout the liquid medium resulting in a thick gel.

The illustration of sol-gel process is shown in Figure 2.15.

R.R. Marlene et al. [56] prepared nanocrystalline TiO_2 wires by the sol-gel method, using TIP (Ti $[OCH\ (CH_3)_2]4$, Aldrich Inc.), isopropanol ($CH_3CHOHCH_3$, Aldrich Inc.), and acetic acid (CH_3COOH, Fluka Inc.).

First, 1 mol of TIP was mixed with 2 mol of isopropanol, after 10 min 4 mol of acetic acid was added. Finally, 20 min later, 2 mol of isopropanol was added and stirred in order to complete 1 h of reaction time. To obtain the xerogel, the sol was dried at 100°C for 24 h.

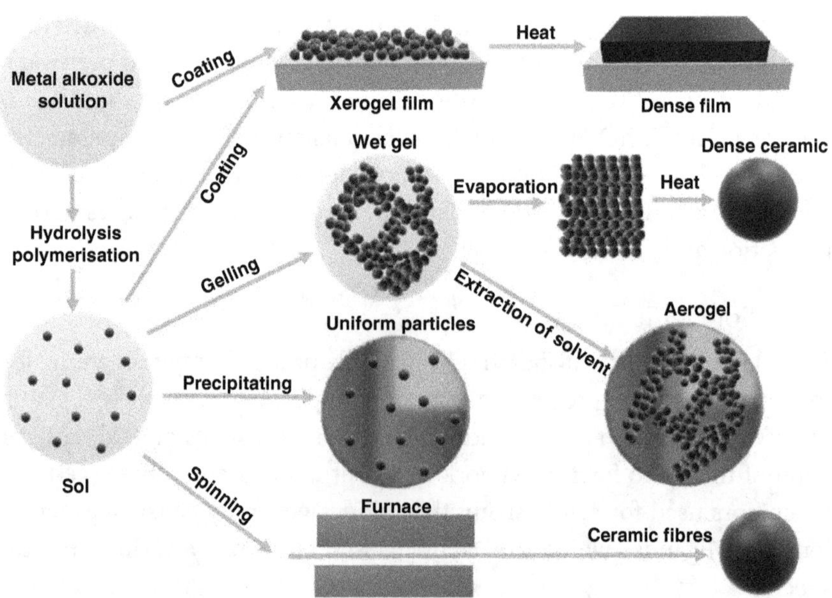

FIGURE 2.15 Sol-gel process.

Finally, the xerogel was calcinated in a conventional furnace at a normal atmosphere at 400, 500, and 600°C for 1 h in order to densify and crystallize the material. After calcination, the samples showed NWs morphology with an average NWs thickness of about 76 nm for all samples. There is no marked difference in size between the materials heat-treated at 400 and 500°C. However, at 600°C, the fibres are longer. This may be related to the remarkable growth of crystallite size and the transformation from anatase to rutile phase.

2.3 SUMMARY

In this chapter, we have presented to our readers, the importance and preparation of 1D metal oxide semiconductor nanostructure in recent research. Although several effective methods have been successfully utilized to design highly efficient metal oxide nanostructures, the biggest obstacle in putting the nanoscale devices into practical application is their high cost. In general, constructing 1D nanostructures needs sophisticated techniques, including lithography-based processes, which are time-consuming and need high-cost instruments. Therefore, the low-cost methods for the large-scale production of metal oxide nanostructures were discussed in detail.

As demonstrated in recent literature, much more novel properties may be expected from heterostructure-based semiconductor nanowires as compared with the corresponding single nanowires. The bandgap-engineered semiconductors can be possible candidates for novel functional optoelectronic devices as they cover a wide range of the spectrum from UV to visible light.

However, the reports on the successful demonstration of wavelength-controllable nanowire-based devices are still limited. Therefore, it is highly needed to develop novel and reliable approaches to synthesize 1D nanostructures with tunable bandgap, by which the wavelength-controlled optoelectronic device could be obtained.

REFERENCES

1. *Handbook of Nanostructured Materials and Nano-technology* (Ed: H. S. Nalwa), Academic Press, New York (2000).
2. M. S. Arnold, P. Avouris, Z. W. Pan and Z. L. Wang, *J. Phys. Chem. B* 107, 6 (2003).
3. E. Comini, G. Faglia, Sberveglier, Z. W. Pan and Z. L. Wang, *Appl. Phys. Lett.* 81, 1869 (2002).

4. A. Bouhelier, R. Bacgelot, J. S. Im, G. P. Wiederrecht, G. Lerondel, S. Kostcheev and P. Royer, *J. Phy, Chem Br* 109, 3195 (2005).
5. S. A. Dong and S. P. Zhou, *Mater. Sci. Eng. B* 140, 153 (2007).
6. S. Sujata, *International Journal of Advanced Research in Engineering and Applied Sciences* 3, 35–40 (2014).
7. E. Monroy, F. Calle, J. L. Pau, E. Munoz, F. Omnes, B. Beaumont and P. Gibart, *J. Cryst. Growth* 230, 537 (2001).
8. M. Razeghi, A. Rogalski, "Semiconductor ultraviolet detectors." J. *Appl. Phys.* 79, 7433 (1996).
9. Y. B. Li, F. D. Valle, M. Simonnet, I. Yamada, J. J. Delaunay, "High-performance UV detector made of ultra-long ZnO bridging nanowires." *Nanotechnology* 20, 045501 (2009).
10. M. Chen, L. F. Hu, J. X. Xu, M. Y. Liao, L. M. Wu, X. S. Fang, "ZnO Hollow-Sphere nanofilm-based high-performance and low-cost photodetector," *Small* 7, 2449–2453 (2011).
11. E. Monroy, F. Omnès, and F. Calle, *Semiconductor Science and Technology*, v. 18, pp. R33–R51 (2003).
12. E. Munoz, E. Monroy, J. L. Pau, F. Calle, F. Omnes and P. Gibart, *J. Phys.: Condens. Matter.* 13, 7115 (2001).
13. Y. Z. Chiou and J. J. Tang, *J. Appl. Phys.* 43, 4146 (2004).
14. M. L. Lee, J. K. Sheu, W. C. Lai, Y. K. Su, S. J. Chang, C. J. Kao, C. J. Tun, M. G. Chen, W. H. Chang, G. C. Chi and J. M. Tsa, *J. Appl. Phys.* 94, 1753 (2003).
15. S. J. Chang, C. L. Yu, C. H. Chen, P. C. Chang and K. C. Huang, *J. Vac. Sci. Technol. A* 24, 3 (2006).
16. N. W. Emanetoglu, J. Zhu, Y. Chen, J. Zhong, Y. Chen, and Y. Lu, *Appl. Phys. Lett.* 85, 3702 (2004).
17. S. Liang, H. Shenga, Y. Liua, Z. Huoa, Y. Lua and H. Shen, *J. Cryst. Growth* 225, 110 (2001).
18. A. Balducci, M. Marinelli, E. Milani, M. E. Morgada, A. Tucciarone, G. Verona-Rinati, M. Angelone and M. Pillon, *Appl. Phys. Lett.* 86, 193509 (2005).
19. H. D. Liu, X. Guo, D. McIntosh and J. C. Campbell, *IEEE Photon. Technol. Lett.* 18, 2508 (2006).
20. K. H. Chang, J. K. Sheu, M. L. Lee, S. J. Tu, C. C. Yang, H. S. Kuo, J. H. Yang and W. C. Lai, *Appl. Phys. Lett.* 97, 013502 (2010).
21. N. S. Allen, M. Edge, J. Verran, J. Stratton, J. Maltby, C. Bygott, *Polymer Degradation and Stability* 93, 1632 (2008).
22. H. Kazuhito, I. Hiroshi, and F. Akira, *Jpn. J. Appl. Phys.* 44, 8269 (2005).
23. S. Hong, A. Han, E.C. Lee, K.-W. Ko, J.-H. Park, H.-J. Song, M.-H. Han, C.-H. Han, *Current Applied Physics*, 15, 574 (2015).

24. S. A. Maier, P. G. Kik, H. A. Atwater, S. Meltzer, E. Harel, B. E. Koel and A. A. G. Requicha, *Nat. Mater.* 2, 229 (2003).
25. K. Zhu, N. R. Neale, A. Miedaner, and A. J. Frank, *Nano Lett.* 7, 69 (2007).
26. A. Karabchevsky, O. Krasnykov, I. Abdulhalim, B. Hadad, A. Goldner, M. Auslender and S. Hava, *Photonics and Nanostructure-Fundamental and Appl.* 7, 170 (2009).
27. R. J. Martín-Palma and A. Lakhtakia, *Nanotechnology: A Crash Course*, SPIE, (2010).
28. A. K. Kar, P. Morrow, X. T. Tang, T.C. Parker, H. Li, J. Y. Dai, M. Shima and G. C. Wang, *Nanotech.* 18, 295702 (2007).
29. H. A. Macleod, *Thin-film Optical Filters*, 3rd ed, Institute of Physics, Bristol, United Kingdom (2001).
30. S. H. Hong, J. Zhu, C. A. Mirkin, *Science* 286, 523 (1999).
31. A. Karabchevsky, O. Krasnykov, I. Abdulhalim, B. Hadad, A. Goldner, M. Auslender and S. Hava, *Photonics and Nanostructure-Fundamental and Appl.* 7, 170 (2009).
32. Z. H. Wu, X. Mei, D. Kim, M. Blumin, H. E. Ruda, J. Q. Liu and K. L. Kavanagh, *Appl. Phys. Lett.* 83, 3368 (2003).
33. Y. P. Leung, Z. Liu and S. K. Hark, *J. Cryst. Growth* 279, 248 (2005).
34. H. W. Kim, S. H. Shim, J. W. Lee, D. Y. Shin, C. H. Wi, B. D. Yoo and C. Lee, *J. Korean Physical Society* 50, 1085 (2007).
35. V. G. Besserguenev, R. J. F. Pereira, M. C. Mateus, I. V. Khmelinskii, R. C. Nicula and E. Burkel, *Int. J. Photoenergy* 5, 99 (2003).
36. R. S. Wagner and W. C. Ellis, *Appl. Phys. Lett.* 4, 89 (1964).
37. Y. Xia, J. A. Rogers, K. E. Paul and G. M. Whitesides, *Chem. Rec.* 99, 1823 (1999).
38. L. Holland, *J. Opt. Soc. Am.* 43, 376 (1953).
39. R. S. Devan, R. A. Patil, J. H. Lin, Y. R. Ma, *Adv. Funct. Mater.* 22, 3326 (2012).
40. A. Kundt, *Annalen der Physik* 27, 59 (1886).
41. F. Kaempf, *Annalen der Physik* 26, 308 (1905).
42. C. Maurin, *Comptes Rendus Hebdomaires des Seances de l'Academie des Sciences* 142, 870 (1906).
43. C. Bergholm, *Annalen der Physik* 43, 1 (1914).
44. J. M. Nieuwenhuizen and H. B. Haanstra, *Philips Tech. Rev.* 27, 87 (1966).
45. R. N. Tait, T. Smy, and M. J. Brett, *Thin Solid Films* 226, 196 (1993).
46. N.O. Young, J. Kowal, *Nature* 183, 104 (1959).
47. P. Chinnamuthu, A. Mondal, N. K. Singh, J. C. Dhar, S. K. Das, K. K. Chattopadhyay, *J. Nanoscience and Nanotechnology* 12, 6445 (2012).
48. J C Dhar, A Mondal, N K Singh and K Chattopadhyay, *J. Appl. Phys.* 113, 174304 (2013).

49. R.S. Wagner, W.C. Ellis, Trans. Matall. *Soc. AIME*, 233, 1053 (1965).
50. By L. J. Lauhon, Mark S. Gudiksen and Charles M. Lieber, *Phil. Trans. R. Soc. Lond. A* 362, 1247 (2004).
51. J-C. Lee, K-S. Park, T-G. Kim, H-J. Choi and Y-M. Sung, *Nanotechnology* 17, 4317 (2006).
52. Jong-Yoon Ha, Brian D. Sosnowchik, Liwei Lin, Dong Heon Kang, and Albert V. Davydov, *Applied Physics Express* 4, 065002 (2011).
53. Lin Yue, Wei Gao, Danyu Zhang, Xuefeng Guo, Weiping Ding, and Yi Chen, *J. Am. Chem. Soc.*, 128, 11042 (2006).
54. Young-Sik Yun, Eun-Hye Kang, In-Sik Yun, Yong-Oock Kim, and Jong-Souk Yeo, *J. Korean Vac. Soc.* 22, 204210 (2013).
55. Yi-Seul Park and Jin Seok Lee, Bull. *Korean Chem. Soc.* 32, 3571 (2011).
56. Marlene Rodr íguez-Reyes, Héctor J. Dorantes-Rosales, *J Sol-Gel Sci Technol* 59, 658 (2011).

CHAPTER 3

Polymer Semiconductors and FET Sensors

Alexy R. Tameev

3.1 INTRODUCTION

In recent years, organic and polymeric materials have been increasingly used to solve several basic and applied physics problems, as well as to design new devices. Research in this direction is stimulated by successes in the synthesis of organic materials with specified well-reproducible parameters, in particular, polymers, due to their diversity, manufacturability, and relative cost-efficiency.

Advanced methods of synthesis or subsequent modification make it possible to vary the conductivity of polymers in a wide range from typical values of dielectrics (below 10^{-10} S/cm) to metallic conductivity values (above 1 S/cm) [1,2]. The importance of the development of this scientific direction was confirmed by the award of the Nobel Prize in Chemistry (2000) to A. Heeger [3], A. McDiarmid [4], and H. Shirakawa [5] for the discovery and creation of conductive polymers.

Conductive polymer materials are divided into two large groups: polymers with ionic conductivity or solid polymer electrolytes and polymers with electronic conductivity. In turn, polymers with electronic conductivity include polymers with a conductivity close in mechanism to the electrical conductivity of metals due to conjugated bonds (this type of polymer compounds is

DOI: 10.1201/9781003464211-3

commonly called "conducting polymers" and "organic metals" [1,2]) and non-conjugated polymers in which electron transfer, in chemical terms, occurs due to redox reactions between adjacent fragments of the polymer chain.

At the macroscopic scale, the current through a semiconductor is determined by the charge carrier concentration n and the carrier drift velocity v, where the latter can be expressed in terms of the mobility μ and the electric field F: $J = env = en\mu F$.

According to this equation, in addition to the field, two parameters n and μ determine the magnitude of the current. However, unlike inorganic semiconductors, in organic semiconductors there is usually no linear relationship between j and F. This is because both the concentration and the mobility of charge carriers in them can depend on the applied field.

As was mentioned earlier, the mobility μ strongly depends on the degree of order and purity of organic and polymer semiconductors and, therefore, largely on the conditions for their thin film preparation. It can reach values of 1–10 cm^2/Vs in molecular crystals but may have lower values such as 10^{-6} cm^2/Vs in disordered materials. The highest mobility values achievable in thin films of organic semiconductors are currently comparable with amorphous Si, the mobility in which is several orders of magnitude lower than in crystalline Si.

The intrinsic concentration of charge carriers n_i in a semiconductor with a band gap E_g and with a density of states N_c (which is equal to the concentration of molecules in an organic semiconductor) is determined by the expression: $n_i = N_c \exp(-E_g/2kT)$.

The use of typical values for organic semiconductors E_g = 2.5 eV and N_c = 10^{21} cm^{-3} leads to hypothetically low intrinsic carrier concentration n_i = 1 cm^{-3} at room temperature. This value is not realized since impurities in the material lead to a much higher concentration in real materials. For comparison: using the corresponding values for Si (E_g = 1.12 eV and N_c = 10^{19} cm^{-3}) gives n_i = 10^{10} cm^{-3}. This evaluation demonstrates that organic semiconductors have extremely low conductivity if they are sufficiently pure. In this chapter, we will consider the physicochemical properties of conductive polymers, with particular attention to charge carrier transport.

3.2 INTRINSIC ELECTRONIC CONDUCTIVITY IN A POLYMER CHAIN

The main difference between polymer dielectrics and polymers with intrinsic electronic conductivity is that the former do not contain conjugated

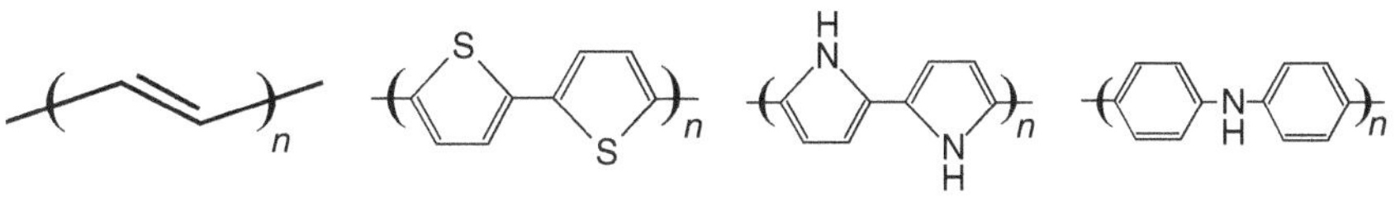

Polyacetylene (10^4) Polythiophene (10^3) Polypyrrole (10^2) Polyaniline (10^2)

FIGURE 3.1 Chemical structure and electrical conductivity (S/cm) of conducting polymers.

FIGURE 3.2 Schematic illustration of the π electrons distribution along the PA chain.

chemical bonds, which are present in the latter. Conducting polymers are characterized by high electron mobility in the π conjugation system. Typical representatives of such polymers are presented in Figure 3.1.

Polymers are macromolecules consisting of repeating molecular "units" – monomeric units (the building block of a polymer) linked by covalent bonds. The band gap in polymers depends on the number of molecular units that form the polymer. In particular, the dependence of the band gap on the number of repeating units has the form:

$$E_G = E_0 + B/N,$$

where N is the number of repeating links; B is a constant. E_g changes insignificantly if the number of links exceeds a certain value N_c. The parameter N_c is called the "number" of conjugations, and it corresponds to the "length" of conjugation l_c, which is determined by the expression $l_c = N_c a$ (where a is the unit cell size).

To depict the structure of polymers, various options are used that show the distribution of electrons along the chain. Considering the formation of the band structure of polyacetylene $(CH)_n$, the different options for polyacetylene are shown in Figure 3.2: (1) π electrons are evenly distributed along the chain. (2) and (3) π electrons are unevenly distributed along the

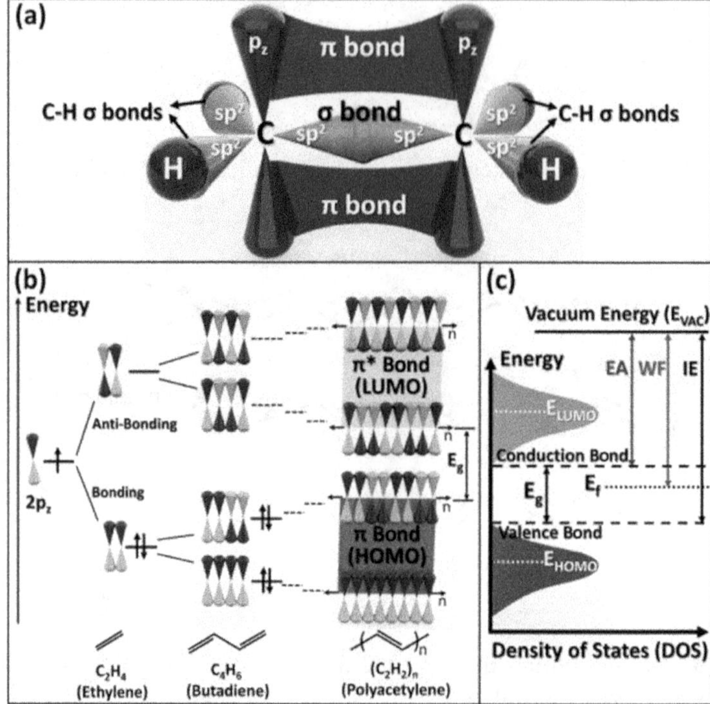

FIGURE 3.3 (a) Schematic illustration of the sigma bond (σ bond) and pi bond (π bond) in a conjugated polymer. Each carbon atom has three σ bonds, two of which are formed with hydrogen and one with the neighbouring carbon atom, and one $2p_z$ orbital that forms π bonds. The π bonds are perpendicular to the σ plane and tend to delocalize along the polymer chain. The illustrated configuration is related to ethylene (CH_2=CH_2) that can also be expanded in a long molecular chain. (b) Schematic illustration of bonding and antibonding of $2p_z$ orbitals for PA, which leads to a splitting of the energy levels to π (filled) and π∗ (empty) bands. (c) Schematic of the energy level diagram that can be considered in both organic and inorganic semiconductors. Evac, Eg, Ef, WF, IE, and EA refer to vacuum level energy, band gap, Fermi energy, work function, ionization energy, and electron affinity, respectively. [Gharahcheshmeh et al. 2020].

chain. Note that the participation of π electrons in the bond reduces the distance between the atoms. The spatial configuration of the chain can be changed by changing its electronic filling. This can happen when (a) an electron is moved to a higher level (e.g., light absorption), (b) an electron is added to the circuit to occupy a higher energy state, or (c) an electron is extracted from a lower energy state (electron or hole injection).

Figure 3.3 shows the schematic illustration of the σ and π bonds between carbon atoms: (a), the energy level splitting of π bonds leads to the

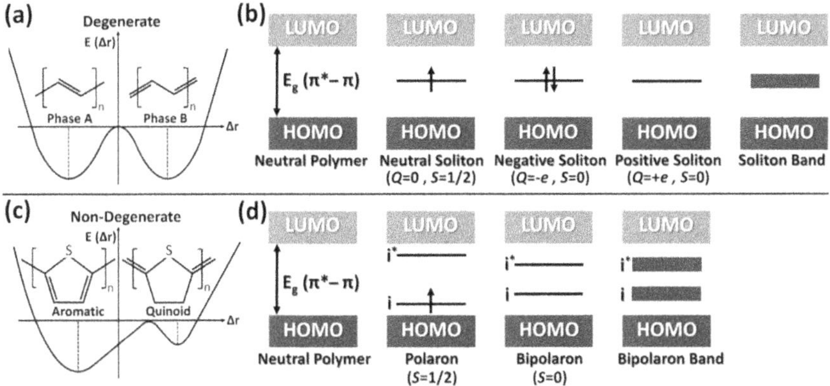

FIGURE 3.4 Potential energy change in two different types of degenerate and non-degenerate ground-state–conjugated polymers. (a) There is no change in potential energy of phase A and phase B with a similar chain on either side of a defect in degenerate ground-state–conjugated polymer and formation of neutral, negative soliton, positive soliton, and soliton band. (b) Potential energy change of aromatic and quinoid in the non-degenerate ground-state–conjugated polymers and formation of polaron, bipolaron, and bipolaron band. The small arrow represents an electron with a spin either up or down. [Gharahcheshmeh et al. 2020].

formation of π (filled) and π∗ (empty) bands for polyacetylene PA (b), and the corresponding energy diagram (c). Molecular π orbitals form HOMO (Highest Occupied Molecular Orbital (valence band)) and LUMO (Lowest Unoccupied Molecular Orbital (conduction band)) states. The band structure of polyacetylene consists of the σ and σ* energy bands formed from a combination of sp^2 hybrid orbitals and the π and π* energy bands formed because of a combination of p_z states of carbon atoms. [6]

The potential energy of a molecular chain is determined by its structure. Figure 3.4a shows the potential energy of a polyacetylene molecule as a function of its spatial organization (in configuration coordinates). As a result of the symmetry of the interatomic bonds of the PA structure, the ground state turns out to be doubly degenerate. Note that the structure of a molecular chain, which is determined by interatomic bonds and leads to the degeneracy of the ground state, is a rare phenomenon. In most cases, a nondegenerate ground state is realized, as shown in Figure 3.4c for polythiophene, with one of the bond configurations being more energetically favourable. The localization of a positive charge in the polymer chain causes a change in the bond configuration, leading to a change in part of the chain from an aromatic to a quinoid bond state. The charge that changes the structure of a molecule is termed polaron. However, since the quinoid part has a higher energy, the molecule will tend to minimize its length. This can happen either by "pushing" the charge (distortion) towards the terminal of the chain, or in the case of a long chain, by the formation of another structural defect (electron capture) that will return the polymer to its more stable (aromatic) bonding state.

Polyacetylene is the simplest conjugated polymer. Each carbon is σ-bonded to two neighbouring carbon atoms and one hydrogen atom. One π electron per carbon remains. The movement of π electrons along the conjugated chain provides further electronic interaction and results in bonding and antibonding orbitals across the polymer [7]. As a result of the combination of p_z orbitals and subsequent coupling between electronic and elastic properties, which is considered as the Peierls' instability, the energy level splitting of π bonds occurs [8–10]. Theoretical calculations performed for a one-dimensional molecular chain with electrons delocalized along the chain in the case of a fixed distance between molecules gave the distribution of the density of electronic states in the $N(E)$ band, which is shown in Figure 3.5a. The shaded area corresponds to the states occupied by charge carriers. There is no band gap, and the polymer should have metallic conductivity. This result was inconsistent with experimental data. Accounting in the theory for different distances between atoms due to different lengths of double and single bonds led to the appearance of a band gap in the conjugated polymer (Figure 3.5b) according to Peierls' theorem. The result showed that the conjugated polymer should possess semiconductor properties.

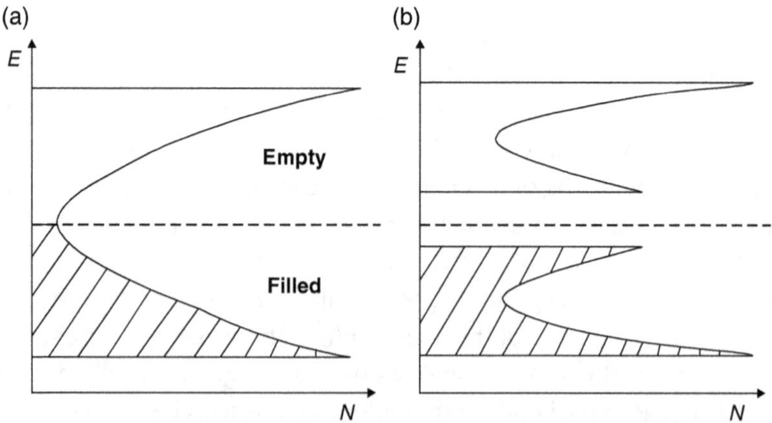

FIGURE 3.5 Density of states for PA in metallic state and dimerized semiconductor state.

Note that the presented energy structure corresponds to one polymer chain. At the same time, the polymeric material consists of many intersecting polymeric chains. Since a certain spread in the energies of the structures of different polymer chains is possible, the presence of "disorder" leads to the appearance of localized states of tails of the density of states near the edges of the conduction and valence bands, like the tails of the density of states in inorganic amorphous semiconductors.

3.3 CHARGE TRANSPORT IN CONJUGATED POLYMERS

Considering the possibility of band transfer of charge carriers in a polymer, the mobility of charge carriers is described by the relationship $\mu = e\tau/m_{eff}$, where e is the electron charge, e is the mean scattering time, and m_{eff} is the effective mass. Assuming an approximation of effective mass is applicable [11], the effective mass is $m_{eff} = \hbar^2/(Wa^2)$, where W is the bandwidth, with $kT << W$, and a is the lattice constant. The mobility of charge carriers within HOMO and LUMO bands in π-conjugated polymers can then be evaluated using $\mu = e\tau Wa^2/\hbar^2$. In accordance with the Ioffe and Regel condition [12], for bands to retain a physical significance, W must be greater than ΔE, with $\Delta E \approx \hbar/\tau$ in accordance with the uncertainty principle. Thus, the following inequality must be satisfied

$$\mu = ETWA^2/\hbar^2 > ET\Delta EA^2/\hbar^2 \approx EA^2/\hbar$$

Experimental measurements have revealed that the typical values of the mobility of charge carriers in such polymers ($\mu > 10^{-4}$ cm^{2}V^{-1}s^{-1}) do not satisfy this inequality [11]. Therefore, in conjugated polymers, the possibility of implementing band transfer of charge carriers is low.

The study of conductivity in polyacetylene (PA) led to the development of a model in which the conductivity in polymers is associated with the movement in the polymer chain of solitons and polarons that can easily move along the molecular chain. These particles differ from electrons and holes in the absence of spin (for a soliton and bipolaron) and a lower velocity of movement.

3.3.1 Polarons

In inorganic crystalline materials, polarons are termed quasiparticles consisting of an electron and the surrounding field of elastic deformation of the polarized lattice. For example, when moving in a lattice, an electron interacts with lattice atoms that have a nonzero effective charge. As a result, a kind of "shell" of crystal deformations forms around the electron. This "shell" is associated with the low mobility of charge carriers in ionic crystals, since such a deformation of the crystal creates a potential well for the electron. The movement of this deformation along with the electron causes a decrease in its kinetic energy.

In disordered polymeric materials, polarons are a segment of a polymer consisting of several monomers with a nonzero total charge, in which the monomers have undergone a change in the structure of electronic bonds. This deformation of the structure of electronic bonds is the reaction of the electronic system of the polymer to an excess or lack of electrons that form π-conjugated bonds. In conjugated polymers, a polaron represents an altered sequence of alternating single and double bonds between atoms of the main polymer chain (see Figure 3.4). The charge carrier is the entire charged segment of the polymer, and when the polaron moves along the polymer chain, the deformation of the electronic system and the structure of electronic bonds is transferred. The polaron transport has an activation nature, since when the polaron moves along the polymer chain, the monomer electron system requires additional energy for the transition to the deformed state and vice versa. During the formation of a polaron, two energy levels arise, associated with the HOMO and LUMO of the perturbed segment of the polymer constituting the polaron. The difference between the energies of the ground states of the unperturbed and perturbed segments of the polymer is the binding energy of the polaron E_{pol}. This binding energy also characterizes the location of the energy levels of the electronic states formed by the formation of a polaron. The value of E_{pol} for bipolarons is greater than that for polarons due to the amplification of the perturbation of the monomer by two charges. The mobility of polarons can be several orders of magnitude lower than the band mobility of electrons and holes and varies within 10^{-7} and 10^{-1} cm^2/Vs. Such low mobilities are explained by the redistribution of heavy atoms accompanying the motion of an electron (hole).

In inorganic semiconductors, polarons of small and large radius are distinguished. The criterion is the ratio of the polaron radius (the size of the atomic redistribution region associated with the presence of an electron or hole) to the lattice constant of the crystal. In organic semiconductors, the change in the positions of atoms occurs within the limits of a resonant structure, molecule, or polymer unit. Thus, polarons in organic semiconductors are of polarons of small radius. Polaron transport is limited mainly by the transition of a charge carrier from one resonant structure to another. At high temperatures (including room temperature), it is most advantageous to overcome the "polaron" barrier due to thermal activation. The probability of thermal activation decreases exponentially with decreasing temperature. Therefore, processes of isoenergetic tunnelling between localized polaron states dominate at low temperatures. With heavy doping of the polymer and, accordingly, with an increase in the concentration of polarons, a polaron band is formed, along which the transfer of polarons occurs. Within such bands, polarons should have a large mass and low mobility. The width of the polaron band, as a rule, is two to three orders of magnitude smaller than the energy band width for electrons and holes and ranges from several meV to several tens of meV. Apart from some highly ordered molecular crystals, band transport of polarons does not occur in polymer materials due to the presence of spatial and energy disorders. Therefore, electrons (holes) in lightly doped polymer semiconductors are localized.

3.3.2 Solitons

Solitons are structural defects that can exist in degenerate polymers, which were discussed earlier. For these polymers, the ground state is degenerate in energy and is a combination of several states with the same energy, differing only in the configuration of the structure of chemical bonds within the monomers. An example of such a polymer is polyacetylene, the structure of which is shown in Figure 3.1 through Figure 3.3. The state at the local maximum of the configuration diagram corresponds to a soliton (Figure 3.3a). Schematic representations of solitons using trans-polyacetylene chains as an example are shown in Figure 3.6. On the top chain, the dot marks the remaining unbound electron, which is not involved in the formation of the π-bond. On the bottom chain, the dotted line shows a possible variant of the delocalization of this electron between the parts of the polymer that are in the ground state. So, it is possible to realize two types of neutral solitons in trans-polyacetylene, as shown in Figure 3.6.

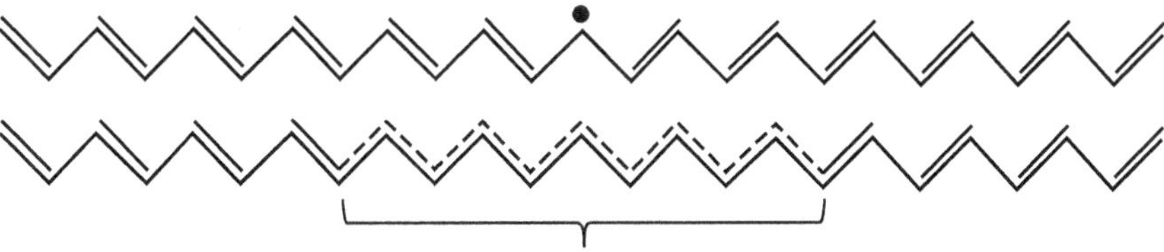

FIGURE 3.6 Scheme of solitons (neutral kinks) in the trans-polyacetylene chain.

In the first type, the electron is localized in one place of the polymer chain. In the second type, the electron is delocalized over a marked region. In polyacetylene, the size of the soliton is seven lattice constants. In both cases, the soliton separates two parts of the polymer chain, which are in different states with the same energy as the ground state. Thus, solitons represent a kind of interface in a quasi-one-dimensional polymer chain between its two segments, in which the monomers are in two different states with an energy equal to the energy of the ground state. In Figure 3.4b, the band structure of the polymer and the position of the soliton level in trans-polyacetylene, located in the middle of the band gap, is schematically shown. A soliton can acquire a positive (negative) charge through a chemical oxidation (reduction) reaction with dopants. In energy terms, this is equivalent to the electron passing from the soliton level to the dopant level (passes to the soliton level from the dopant level). The interaction between charged solitons leads to the formation of a soliton band of localized electronic states. This band becomes wider as the doping level increases since the concentration of solitons increases and, consequently, the interaction between charged solitons increases. The soliton transport is an activation process since moving this structural defect from one monomer to another requires an energy equal to the height of the potential barrier, which, in the case of a doubly degenerate ground state (Figure 3.4a), separates the two ground states of the polymer.

Thus, for a fixed value of some parameters of polymers (e.g., those that determine their optical properties), it is sufficient to have a molecule length equal to or somewhat greater than the conjugation length. However, this does not apply to the charge carrier transport properties that determine the conductivity. Here it is necessary to consider not only intramolecular transfer but also intermolecular charge transfer. Therefore, the conductivity will be greater for longer polymer molecules because less intermolecular hopping between linear molecules is required.

3.4 CHARGE TRANSPORT IN NON-CONJUGATED POLYMERS

The charge transport mechanism is occurred by hopping between adjacent transport centres (TCs) which from distributed density of sites in energy [13]. It is a matter of general experience that the most effective TCs are of (1) segments of the chain containing σ- or π-conjugated bonds and (2) aromatic functionalities (amino-phenyl, hydrazone, carbazolyl). TCs are located on the main chain and on the pendant moiety of a polymer as

(a) 

(b) 

(c)

FIGURE 3.7 Schematic representation of charge-transport polymers. Green areas indicate a segment of conjugated bonds and charge transporting centres (TCs). Polymers possess TCs formed by (a) delocalized π-conjugated bonds in the main chain, (b) localized states either in the main chain or (c) in pendant groups.

shown in Figure 3.7a, 3.7b, and 3.7c, respectively. In conductive polymers considered above, segments of π-conjugated bonds in the main chain serve as TCs.

The conductivity of all such polymers is limited by intermolecular (interchain) charge transfer, which requires overcoming a spatial and energy barrier. In polymer solid films which are of a type of disordered non-crystalline medium, charge carriers moving by hopping between neighbour TCs experience capture and scattering on defects at each hopping event; therefore, the mobility is low $\mu << 1\ cm^2V^{-1}s^{-1}$. Therefore, the charge transport in a polymer medium is an activation process in which lattice vibrations stimulate either tunnelling or over-barrier hopping between TCs, that is, it is the phonon-assisted transport.

Tunnelling depends on the distance r between the TCs and is determined by the overlap integral of the corresponding wave functions:

$$I(r) = I_0 \exp\left(-\frac{2r}{\alpha}\right),$$

where α is the characteristic radius of the wave function. The rate of intermolecular electron transfer v_{ij} from i to j TC can be represented as an expression proposed by Marcus [14,15]:

$$v_{ij} = \frac{|I_{ij}|}{\hbar}\sqrt{\frac{\pi}{\lambda kT}} exp\left(-\frac{\left(\Delta G_{ij} + \lambda\right)^2}{4\lambda kT}\right) \tag{3.1}$$

where λ is the reorganization energy, which is the sum of the internal reorganization energy of the molecule and the polarization of the medium;

kT is the thermal energy; and ΔG_{ij} is the difference in the energy of states i and j. Thus, in (3.1), polaronic autolocalization and energy disorder are considered. Nevertheless, it is often difficult to distinguish their influence, since both the polarization of the medium and the energy disorder have the same effect on the parameters measured in experiments.

The rate of charge carrier transition between TCs is represented by the Miller-Abrahams expression [16]:

$$\nu(r_{ij}) = \nu_0 \exp - \left(\frac{2r_{ij}}{a} \right) \begin{cases} \exp\left(-\frac{\Delta E_{ij}}{kT} \right), \Delta E_{ij} > 0 \ (i \to j, \text{ upward}) \\ 1, \qquad \Delta E_{ij} \leq 0 \ (i \to j, \text{downward}) \end{cases} \tag{3.2}$$

where ν_0 is the electronic frequency factor (or the so-called release attempt frequency). The difference in energy between states i and j considers only the energy disorder.

Based on the Miller-Abrahams approximation, Bässler proposed a model in which he proceeded from the fact that localized TCs are distributed over energy [13, 17]. Since the absorption spectra of polymers have a Gaussian shape, the distribution of the density of states was assumed to be Gaussian. The motion of charge carriers is considered as jumps (tunnelling transitions involving phonons) inside and between chains. If a charge carrier is localized (due to disorder or self-localization), then its movement from one TC to another requires lattice vibrations. So, during hopping transport, the mobility of charge carriers increases with temperature.

Fluctuations in the distance between TCs, which are reflected in the overlap integral, make a significant contribution to the spatial disorder. In turn, potential fluctuations or relaxation (reorganization) of molecules naturally make TCs non-equivalent in intermolecular electron transfer. The nature of the distribution of the density of states (DoS) imposes its own characteristics on the transport of charge carriers.

An approach for determining the energy distribution of DoS based on the analysis of temperature dependence of charge carrier mobility together with absorption coefficient spectral dependence was proposed [18]. A constant photocurrent method (CPM) [19, 20] was used to reveal optical transitions leading to nonequilibrium charge-carrier generation. A photoconductive PPQ-DBT co-polymer was chosen as a material for the study (Figure 3.8). A polyphenylquinoline derivative PPQ, a part of the co-polymer, includes triphenylamine moiety (TPA) and has an optical

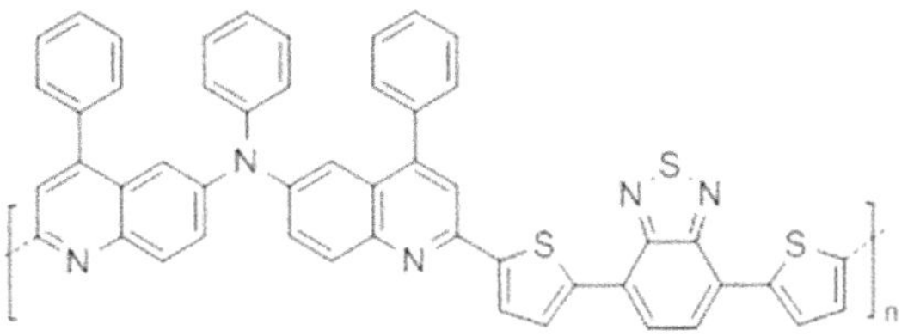

FIGURE 3.8 Chemical structure of a polyphenylquinoline polymer with a phenylamine bridging group and a 2,1,3-benzothiadiazole molecule (PPQ-DBT).

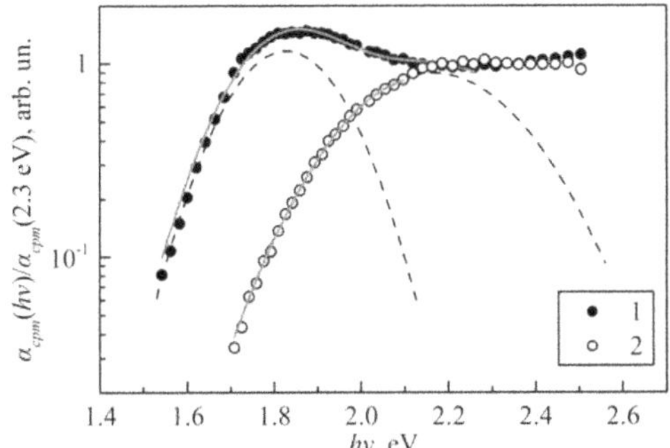

FIGURE 3.9 The spectral dependences of the CPM absorption coefficient α_{cpm} (hν) normalized to the value at hν = 2.3 eV for pristine (1) and annealed at 150°C for 30 min in vacuum; (2) PPQ-DBT film. Red lines show the experimental dependences approximated by Gauss functions. The relevant Gaussians are shown by dashed lines.

band gap of 2.8 eV [21]. It was shown that a composite of dithienyl-2,1,3-benzothiadiazole (DBT) molecules embedded in the PPQ matrix possess photoconductivity and a narrow band gap of 1.8 eV [22].

The spectral dependences of the absorption coefficient $\alpha_{cpm}(h\nu)$ measured at CPM conditions are shown in Figure 3.9. The $\alpha_{cpm}(h\nu)$ dependence of the pristine film is approximated well by a superposition of the two Gaussian functions (Eq. 3.3a) and the α cpm (hν) dependence of the annealed film is approximated by one Gaussian function (Eq. 3.3b):

$$\alpha_{cpm}(h\nu) \sim A\cdot\exp\left(-\frac{\left(h\nu - E_{g1}^{*}\right)^2}{2w_1^2}\right) + B\cdot\exp\left(-\frac{\left(h\nu - E_{g2}^{*}\right)^2}{2w_2^2}\right), \tag{3.3a}$$

$$\alpha_{cpm}(h\nu) \sim C\cdot\exp\left(-\frac{\left(h\nu - E_{g}^{*}\right)^2}{2w^2}\right), \tag{3.3b}$$

where A, B, and C are constants, $E_{g1}^* = (1.83 \pm 0.01)$ eV, $E_{g2}^* = (2.17 \pm 0.04)$ eV, $w_1 = (0.123 \pm 0.013)$ eV, $w_2 = (0.225 \pm 0.085)$ eV, $E_g^* = (2.17 \pm 0.01)$ eV, and $w = (0.176 \pm 0.003)$ eV.

As the absorption coefficient $\alpha_{cpm}(h\nu)$ is determined by the density distribution of the "initial" $N_i(E)$ and "final" $N_f(E)$ electronic states involved in optical transitions and photoconductivity, the spectral dependence of $\alpha_{cpm}(h\nu)$ relates with these parameters by the following expression [20]:

$$\alpha(h\nu) \sim \int \frac{N_i(E) N_f(h\nu + E) |M^2|}{h\nu} dE \tag{3.4}$$

where M is the transition matrix element, which is usually considered to be weak dependent on energy E. The initial and final states are considered to correspond to the electronic states of the highest occupied molecular orbital (HOMO) and the lowest unoccupied molecular orbital (LUMO), respectively. According to the results of the analysis of absorption coefficient spectral dependence and expression (3.4), absorption Gaussians (Figure 3.9) arise due to optical transitions between the initial and final electronic states having Gauss distribution over energy. Moreover, it follows from (3.4) that the full-width at-half-maximum (FWHM) observed for each absorption spectrum w is related to the FWHM of the Gaussian distributions of the initial ω_i and final ω_f states as $w^2 = \omega_i^2 + \omega_f^2$. The obtained result agrees with the generally accepted fact that for polymer and organic disordered materials, the DoS distribution is a superposition of Gaussian functions [13, 17, 23–26].

3.5 FET DEVICES BASED ON POLYMER SEMICONDUCTORS

Conducting polymers are attractive materials for both existing and emerging technologies due to the discussed experimental and theoretical findings together with advantages over other standard conductors and semiconductors, such as their lightweight, low cost, and versatility. Some examples of potential applications include rechargeable batteries, electrochromic displays and smart windows, light emitting diodes (LEDs), photovoltaic cells, toxic waste cleanup, sensors, field effect transistors (FET), electromagnetic interference shielding, and so on. Among these, we will consider FETs, which after the use of organic photoreceptors for xerography and LEDs are the next application of organic semiconductors with the potential of becoming commercially significant at least in printed electronics.

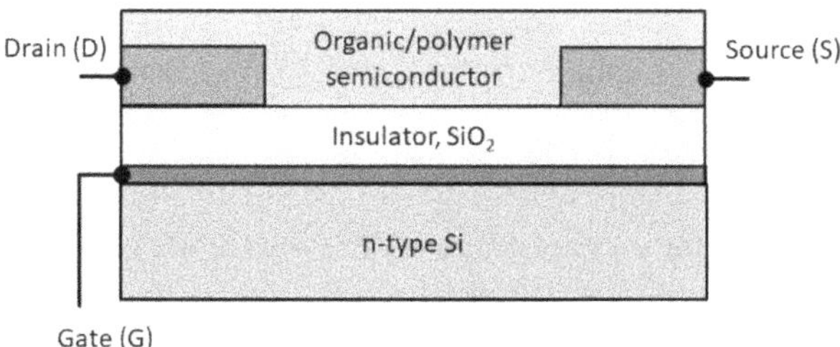

FIGURE 3.10 Cross sectional scheme of a typical OFET with a bottom gate.

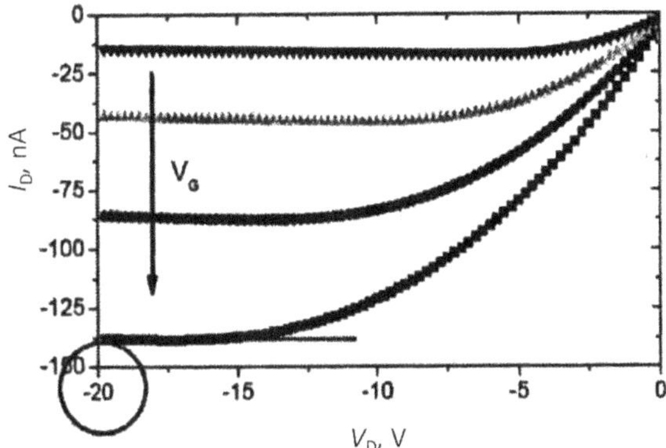

FIGURE 3.11 A typical drain-source current (I_D) vs drain-source voltage (V_D) characteristics at various V_G.

The structure of an organic FET (OFET) is depicted in Figure 3.10. In this case, OFET is considered, in which the charge transport between the source (S) and the current in the drain-source circuit can be changed by changing the voltage applied to the gate (G).

The polymer semiconductor is deposited to the top of the device from a solution. The current–voltage characteristics of such OFETs are typical for the structure of FETs and correspond to the standard operating parameters of FETs based on inorganic semiconductors. Figure 3.11 reveals a typical *I-V* dependence for "p channel" at constant values of the voltage applied to the gate V_G. When the gate voltage exceeds a threshold value (V_T), charge carriers are accumulated (or depleted) in the region at the interface between the insulator and the semiconductor, and form (or remove) a "channel," which allows the current to flow between source and drain. At small values of V_D, I_D increases linearly with increasing V_D, this range is described by the expression [27]:

$$I_D = \frac{W\mu C_i}{L}(V_G - V_T)V_D,,$$

where W is the width of the electrodes, C_i is the insulator capacitance per unit area, L is the channel length, μ is the (average) carrier mobility, and V_T is the threshold voltage. When the value of V_D becomes greater than V_G, then I_D tends to saturate (saturation mode) due to the cutoff of the accumulation layer, and this mode is described by the following relationship [27]:

$$I_D = \frac{W\mu C_i}{2L}(U_G - U_T)^2$$

OFET devices have good potential for sensor applications because the polymer functional moieties are designed to identify the detected substances. Moreover, both the semiconductor layer and gate dielectric layer can be used as the test site, and the changes in μ, I_D, or V_T can be used as detection signals. OFET-based sensors include gas, liquid-phase chemical, temperature, pressure sensors, and so on.

In this regard, research on the use of conductive polymers as active material in field-effect transistor devices was started more than 20 years ago [28, 29]; dedoped or doped conjugated polymers such as trans-polyacetylene [30], polythiophene [31, 32], poly(3-alkylthiophenes) [33], poly(N-methylpyrrole) [34], and polyaniline [35] were demonstrated as suitable materials for OFET architecture.

Among the various OFET-based sensors, gas sensors are widely studied and the detection of NH_3, NO_2, H_2S, SO_2, or other gases has been achieved [36, 37]. Usually, gas detection mainly depends on the interaction between gas molecules and semiconductor molecules. Obviously, a single molecular-level thin active layer can significantly improve the sensitivity. For example, Zhang et al. developed a method (on-the-fly-dispensing-spin-coating) for preparing organic layers of 2–10 nm in thickness and achieved high sensitivity and a fast recovery rate for 10 ppm of NH_3 [38].

The first well-balanced ambipolar OFET array via conventional photolithography based on diketopyrrolopyrrole (DPP)-selenophene copolymers with SiC4-terminated side chains (PTDPPSe-SiC4) semiconductor and graphene electrodes was reported (Figure 3.12) [39]. The devices with graphene electrodes showed ambipolar performance with hole and electron mobilities of 1.43 and 0.37 $cm^2V^{-1}s^{-1}$, respectively. The OFET arrays

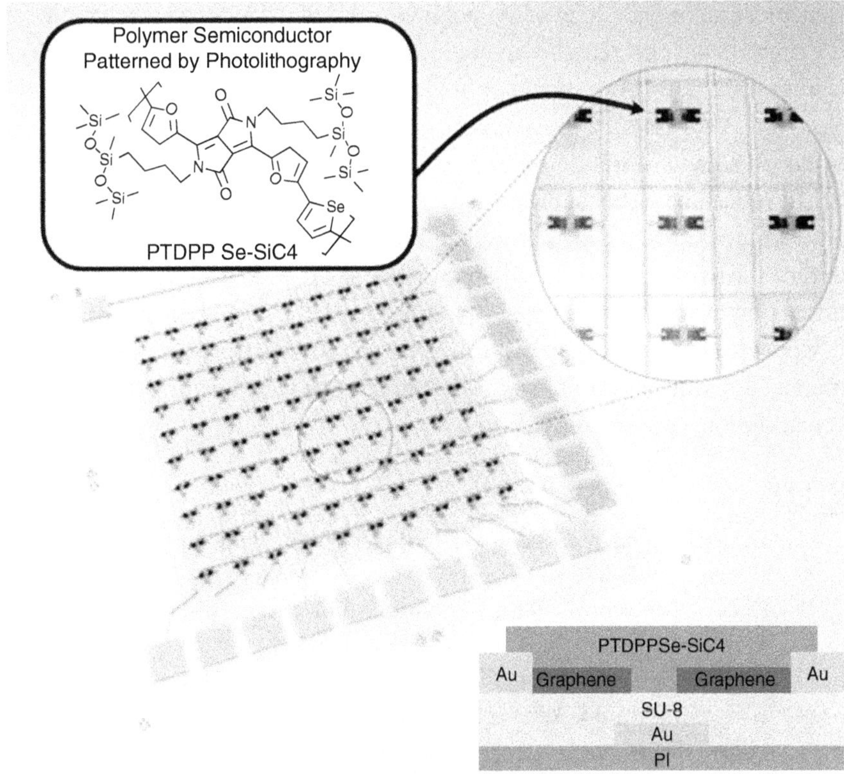

FIGURE 3.12 Ambipolar OFET arrays based on PTDPPSe-SiC4 polymer semiconductor and graphene electrodes [39]. Graphical Abstract.

can detect acetone vapours by decreasing the I_D current with a response time of 0.88 s.

For further study of OFET-based sensors and biosensors and the use of polymer semiconductors in them, a few excellent reviews are useful [36, 40–42].

3.6 CONCLUSION

The π-orbitals play the main role in the formation of the semiconductor properties of polymers. They form low-energy states of the polymer, resulting in the formation of a HOMO and LUMO energy structure with small values of the gap. This makes it possible to dope the material and inject charge carriers into it from the contacts. The conductivity of all considered polymers is limited by intermolecular (interchain) charge transfer, which requires overcoming a spatial and energy barrier. In this regard, the effect of the supramolecular structure and morphology of the polymer solid films on the characteristics of the charge transport remains an urgent problem. The solution to this problem is important both from the point of view of obtaining new knowledge about electronic processes in disordered media, since charge-transport processes occur on a nano- and micrometre scale, which coincides with the characteristic dimensions of the supramolecular structure of typical polymer systems and compositions and from the point of view of optimization development of film electronic devices based on polymer semiconductors. Although functional polymer FETs have shown great progress, the stability of materials and devices remains a challenge. This directly determines the possibility of commercial application of organic integrated circuits with low noise, low power consumption, and long service life.

REFERENCES

1. G. Inzelt. *Conducting Polymers*. Springer Berlin, Heidelberg. (2012) 309 p.
2. *Conjugated Polymers: Properties, Processing, and Applications*, 4th ed. (Eds.: J.R. Reynolds, B.C. Thompson, and T.A. Skotheim). CRC Press. (2019).
3. A.J. Heeger, *Angew Chem Int. Ed.* **40**, 2591 (2001).
4. A.G. MacDiarmid, *Angew Chem Int. Ed.* **40**, 2581 (2001).
5. H. Shirakawa, *Angew Chem Int. Ed.* **40**, 2574 (2001).
6. M. Heydari Gharahcheshmeh, K.K. Gleason, *Materials Today Adv.* **8,** 100086 (2020).
7. The WSPC Reference on Organic Electronics: Organic Semiconductors. Volume 1: Basic Concepts. (Eds.: J.-L. Bredas and S. R. Marder) World Scientific. 436 p. (2016).
8. S.A. Brazovskii, N.N. Kirova, S.I. Matveenko. *Sov. Phys. JETP*, **59** (2), 434 (1984).
9. A. J. Epstein. Insulator-metal transition and transition and metallic state in conducting polymers. In: *Handbook of Conducting Polymers*, 3rd ed. (Eds.: T. A. Skotheim and J. R. Reynolds) 15-1–15-75. Boca Raton, FL: CRC Press. (2007).
10. M. Bharti, A. Singh, S. Samanta, D.K. Aswal. *Prog. Mater. Sci.* **93**, 270–310 (2018).
11. A. Moliton and R. C. *Hiorns. Polym. Intern.* **53**, 1397–1412 (2004).
12. A.F. Ioffe and A.R. Regel, *Noncrystalline, Amorphous and Liquid Electronic Semiconductors, Progress in Semiconductors*, Vol 4, Heywood and Co Ltd, London (1960).
13. A Köhler, H Bässler. *Electronic Processes in Organic Semiconductors: An Introduction.* Wiley-VCH Verlag GmbH & Co. KGaA, Weinheim (2015).

14. R.A. Marcus. *J. Chem. Phys.* **24**, 966–978 (1956).
15. R.A. Marcus. *Rev. Modern Physics.* **65**, 599–610 (1993).
16. A. Miller and E. Abrahams. *Phys. Rev.* **120**, 745–755 (1960).
17. H. Bässler H. *Physica status solidi* (b). **175**, 15–56 (1993).
18. S. R. Saitov, D. V. Amasev, A. R. Tameev, A. G. Kazanskii. *Organic Electronics.* **86**, 105889 (2020).
19. M. Vaněček, J. Kočka, J. Stuchlík, Z. Kožíšek, O. Štika, A. Tříska. Sol. Energy Mater. **8**, 411–423 (1983).
20. V.V. Malov, A.R. Tameev, S.V. Novikov, M.V. Khenkin, A.G. Kazanskii, A.V. Vannikov. *Semiconductors.* **50**, 482–486 (2016).
21. V.M. Svetlichnyi, E.L. Aleksandrova, L.A. Myagkova, N.V. Matyushina, T.N. Nekrasova, R.Yu. Smyslov, A.R. Tameev, S.N. Stepanenko, A.V. Vannikov, and V.V. Kudryavtsev. *Semiconductors.* **45**, 1339–1345 (2011).
22. E. L. Aleksandrova, V. M. Svetlichnyi, N. V. Matyushina, L. A. Myagkova, V. V. Kudryavtsev, and A. R. Tameev. *Semiconductors.* **48**, 1481–1484 (2014).
23. M. Pope, C.E. Swenberg, *Electronic Processes in Organic Crystals and Polymers*, Oxford University Press, New York, 1999.
24. S.D. Baranovskii. *Phys. Status Solidi* **215**, 1700676 (2018).
25. A. Foertig, A. Baumann, D. Rauh, V. Dyakonov, C. Deibel. *Appl. Phys. Lett.* **95**, 052104 (2009).
26. A.J. Campbell, D.D.C. Bradley, D.G. Lidzey. *J. Appl. Phys.* **82**, 6326–6342 (1997).
27. G. Horowitz, R. Hajlaoui, H. Bouchriha, R. Bourguiga, M.E. Hajlaoui. *Adv. Materials.* **10**, 923–927 (1998).
28. F. Cacialli. *Phil. Trans. R. Soc. Lond.* A **358**, 173–192 (2000).
29. M. S. Freund and B. A. Deore. *Self-Doped Conducting Polymers.* Wiley and Sons, 338 p. (2007).
30. J. H. Burroughes, C. A. Jones, R. H. Friend. *Nature*, **335**, 137 (1988).
31. A. Tsumura, H. Koezuka, T. Ando. *Appl. Phys. Lett.* **49**, 1210 (1986).
32. L. Torsi, A. Tafuri, N. Cioffi, M.C. Gallazzi, A. Sassella, L. Sabbatini, P.G. Zambonin. *Sens. Actuators B.* **93**, 257–262 (2003).
33. J. Paloheimo, H. Stubb, P. Ylilahti, P. Kuivalainen. *Synth. Metals.* **41**, 563 (1991).
34. A. Tsumura, H. Koezuka, S. Tsunoda, T. Ando. *Chemistry Letters.* **15**, 863–866 (1986).
35. R. K. Yuan, S. C. Yang, H. Yuan, R. L. Jiang,H. Z. Qian, D. C. Gui. *Synth. Metals.* **41**, 727 (1991).
36. Y.H. Lee, M. Jang, M.Y. Lee, O.Y. Kweon, and J.H. Oh. *Chem* **3**, 724–763 (2017).
37. K. Amer, A. M. Elshaer, M. Anas, S. Ebrahim. *J. Materials Science: Materials in Electronics.* **30**, 391–400 (2019).

38. F. Zhang, C.A. Di, N. Berdunov, Y. Hu, X. Gao, Q. Meng, H. Sirringhaus, and D. Zhu. *Adv. Mater.* **25**, 1401–1407 (2013).
39. E.K. Lee, C.H. Park, J. Lee, H.R. Lee, C. Yang, and J.H. Oh. *Adv. Mater.* **29**, 1605282. (2017).
40. D. Elkington, N. Cooling, W. Belcher, P. C. Dastoor and X. Zhou. *Electronics*, **3**, 234–254 (2014).
41. J. Yang, Z. Zhao, S. Wang, Y. Guo, and Y. Liu. *Chem* **4**, 2748–2785 (2018).
42. S. Hong, M. Wu, Y. Hong, Y. Jeong, G. Jung, W. Shin, J. Park, D. Kim, D. Jang, J.-H. Lee. *Sensors and Actuators B: Chemical.* **330**, 129240 (2021).

CHAPTER 4

Fabrication of Gas Sensor Using Metal Oxide Semiconductor Nanostructures

Bijit Choudhuri and Parul Raturi

4.1 INTRODUCTION

Humans receive information from their surroundings by using their sensory organs. Based on the receptor cells' organization associated with these organs, the human can distinguish a wide range of signals with variable intensity. However, many signals are present beyond our sensing range, which may have technological importance. For example, rapid industrial growth results in the emission of different toxic inorganic and volatile organic compound (VOC) gases in the environment outside the olfactory limit of humans [1]. Some of them, mainly nitrogen oxides (NO, NO_2), affect cardiac action and neuron operation, giving rise to neurodegenerative disorders and lung and respiratory diseases [2,3]. The threshold limit value (TLV) is the maximum amount of gas under whose exposure a human is allowed to work for 8 hours without developing any illness. The TLV for NO, NO_2, and CO are 25, 3, and 50 ppm, respectively. These data reveal that the mentioned gases can harm the human body even at a minor concentration. Air pollution leads to ~7 million deaths a year and costs an amount of USD 5 trillion to the global economy due to deterioration in life standards and productivity [4]. These gases are also responsible for global warming, acid rain, photochemical smog, and so on [5]. Thus, the detection of these gases

DOI: 10.1201/9781003464211-4

is very much essential to alleviate the health condition of human civilization. In a gas sensing device, gas analyte molecules' physisorption/chemisorption changes the active material's electrical properties. This change in the characteristics can be mapped with the properties of gas molecules. Apart from toxic gas detection, gas sensors are also used to implement electronic nose, automobile industry, space applications, petroleum refinery, combustion and emission monitoring, natural resource exploration, explosive detection, and so forth [6,7]. Another gas sensor application domain is breath analyser design in which sensing the presence of ketone in human breath can predict the symptoms of diabetes. These motivations lead to an extensive investigation to explore the efficient novel gas sensing devices, as can be seen from the bar diagram in Figure 4.1.

The recent decades have observed a remarkable growth in nanotechnology research due to the salient features of nanomaterials. Due to the dimension constraints in their basic building block, the nanostructures exhibit various exciting properties such as quantum confinement, surface plasmon resonance, high surface-to-volume ratio (SVR), electronic bandgap tuning, and so on [8, 9]. For a given mass of material, a high SVR enables the active material to adsorb more gas molecules. Moreover, the material's geometry and size drastically influence charge carriers' movement in the semiconductor materials. The metal oxide semiconductors are frequently used to realize highly efficient gas sensors for their high sensitivity, high stability, high reproducibility, fast response, superior recovery

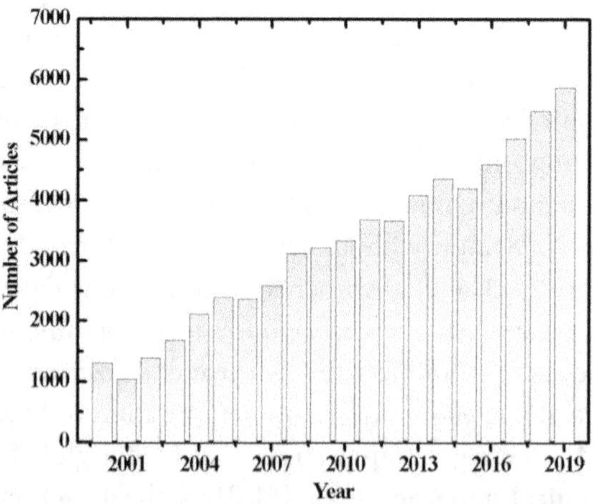

FIGURE 4.1 Bar diagram representing the number of articles with gas sensor keyword published in recent years. (Statistic resource: Scopus).

kinetics, long lifetime, low power consumption, lower processing cost, opportunities for miniaturization, ability to be operated in harsh environments, and so on [2,7,10]. Among these oxide semiconductors, ZnO, SnO_2, and TiO_2 are being investigated extensively due to low cost, high electron mobility, inherent oxygen defect, intrinsic n-type conductivity, outstanding electronic properties, and excellent thermal and chemical stability [11–13]. However, the oxide nanostructures require a high temperature (100–400°C) environment for efficient sensing. This event incorporates additional microheater units in the device and power consumption, sensor reliability, and lifetime of the device degrades [14, 10]. Different approaches such as UV irradiation, surface modification, and impurity addition are implemented to reduce the gas sensor's working temperature [7]. Although a solid-state gas sensor is not an expensive one, the fabrication of its driving and interfacing circuit for CMOS integration needs to be carried out using a discrete PCB component that shoots its price in order [15]. Hence, this chapter is primarily categorized in the basic principle of gas sensing, the improvement in figures of merits (FoMs) of the gas sensor using various techniques. This chapter summarizes the current research associated with selecting novel and efficient materials to develop gas sensors.

4.2 WORKING PRINCIPLES OF GAS SENSORS

The sophisticated and expensive fabrication processes associated with silicon synthesis allow us to obtain an intrinsic single crystalline wafer. The silicon crystal properties can be tuned by intentional replacing of host atom from the lattice by introducing positive or negative impurity. During the growth of the metal oxide semiconductors, there may be insufficient oxygen (metal) environment. In such a scenario, the excess metal (oxygen) can be placed in the lattice's interstitial sites. The event occurs by forming cations (anions) in which they donate electrons (holes). These excess electrons (holes) are primary contributors to inherent n-type (p-type) conductivity in oxide semiconductors. ZnO, SnO_2, TiO_2, Fe_2O_3, WO_3, In_2O_3, and MoO_3 are some of the important n-type oxide semiconductors (NOS) while the NiO, Co_3O_4, Mn_3O_4, Cu_xO, (x=1,2) and Cr_2O_3 are some of the popular p-type oxide semiconductors (POS). Although the response of the POS sensors is lower than the NOS sensors' response, they have the advantages of being less humidity-dependent, high catalytic reaction, and so on [16]. As per the data available with Scopus on 20 May 2020, the count of reports about POS gas sensors is 2212 compared to the 13843 number of articles on NOS gas sensors, as shown in Figure 4.2.

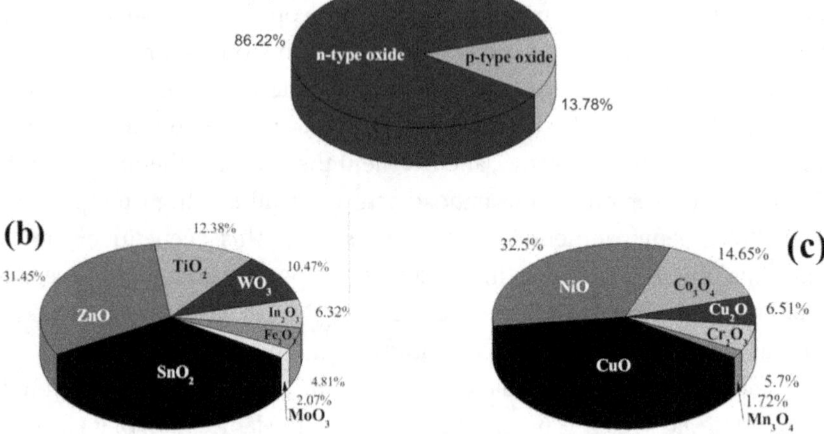

FIGURE 4.2 Studies on NOS and POS gas sensors. (a) Pie chart depicting the comparison of publications on gas sensors based on n-type and p-type oxide semiconductors. (b) Pie chart showing the distribution of different NOS for gas sensors in terms of article number. (c) Pie chart showing article distribution on different NOS-based gas sensors. The general keyword used for the search was the M_xO_y gas sensor where M=metal formula, x and y are the valencies of oxygen and the metal. (The data are collected from Scopus on 20.05.2020.)

The gas sensors are classified into two broad categories depending on their sensing mechanism [18]. Under the exposure of gas molecules, one group of sensors exhibits modulation in their electrical properties while the other sensors show variation in their optical and acoustic properties. This chapter will focus on the electric effect-based gas sensors.

When NOSs are exposed to a high-temperature environment (100–400°C), the oxygen molecules at the exterior are adsorbed and occupy oxygen vacancies to create anions (O^-, O_2^-, O^{2-}) by taking the free electrons from the conduction band [19]. Thus an electron depletion layer (EDL) originates at the exterior. Therefore, a core-shell structure morphology with a conducting core and highly resistive shell region is formed. In an analogous manner, the hole accumulation layer (HAL) is created by the adsorption of oxygen in POS. In this case, a highly resistive core and conducting shell are formed [20].

$$O_2(g) \rightarrow O_2(adsorbed) \tag{4.1}$$

$$O_2(adsorbed) + e^- \xrightarrow{T<100^\circ C} O_2^-(adsorbed) \tag{4.2}$$

$$O_2\left(adsorbed\right)+2e^- \overset{100^\circ C<T<300^\circ C}{\rightarrow} 2O^-\left(adsorbed\right) \tag{4.3}$$

$$O^-\left(adsorbed\right)+e^- \overset{300^\circ C<T}{\rightarrow} O^{2-}\left(adsorbed\right) \tag{4.4}$$

The ionosorption of oxygen results in Fermi-level pinning at interface, and a band bending is observed in the energy band diagram to accommodate this depletion of electrons/accumulation of holes. The formation of the EDL and HAL and their corresponding band diagrams are shown in Figure 4.3.

When an NOS material adsorbs reducing gas molecules (CO, HCHO, H_2S, NH_3), the oxygen anions in the EDL region oxidize the gas molecules, leading to the injection of electrons into the core. This event due to the potential barrier lowering enhances the electron concentration, leading to an increase in the sensors' conductivity. Similarly, reducing gases with anions in the HAL region leads to a lower hole concentration and subsequent rise in sensor resistance. This behaviour is summarized in Table 4.1.

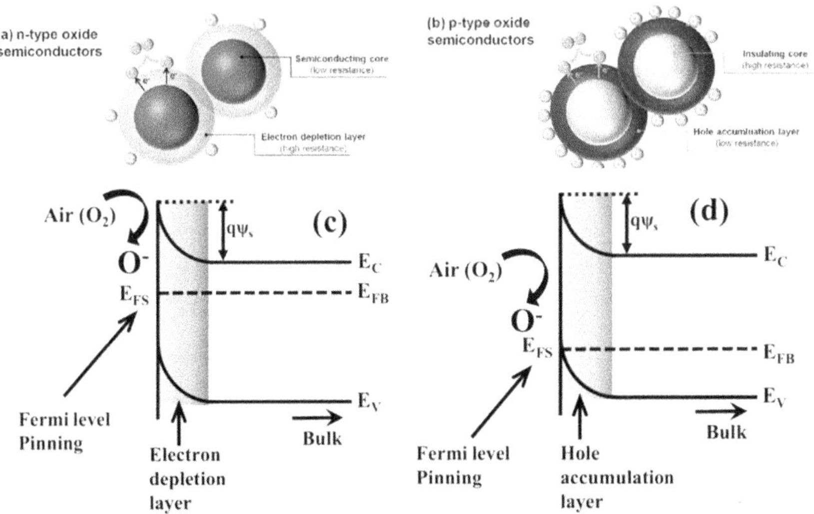

FIGURE 4.3 Formation of electronic core-shell structures in (a) NOS and (b) POS. [Reprinted from [20]. Highly sensitive and selective gas sensors using p-type oxide semiconductors: Overview. Hyo-Joong Kim, Jong-Heun Lee, Sensors and Actuators B: Chemical, 192:607–627, 2014 with permission from Elsevier.] Energy band diagram after the creation of (c) EDL in NOS, (d) HAL in POS. Ec: conduction band minima, Ev: valence band maxima, E_{FB}: bulk Fermi level, E_{FS}: surface to bulk barrier potential.

TABLE 4.1 Summarizes the Change in Resistance in POS and NOS under Gas Exposure

		Underexposure of Reducing Gas		Underexposure of Oxidizing Gas	
Material	**Nature of Shell**	**Majority carrier concentration**	**Resistance**	**Majority carrier concentration**	**Resistance**
NOS	Electron depletion layer	Increase	Decrease	Decrease	Increase
POS	Hole accumulation layer	Decrease	Increase	Increase	Decrease

This modulation in resistance varies proportionately with the amount of analyte gas molecules adsorbed by the semiconductor surface and injected/ejected electrons concentration. The reverse phenomenon is observed for oxidizing gases.

For reducing gas [21]:

$$2CO + O_2^- \left(adsorbed\right) \rightarrow 2CO_2 + e^- \tag{4.5}$$

$$CO + O^- \left(adsorbed\right) \rightarrow CO_2 + e^- \tag{4.6}$$

$$CO + O^{2-} \left(adsorbed\right) \rightarrow CO_2 + 2e^- \tag{4.7}$$

For oxidizing gas [22]:

$$NO_2 + e^- \rightarrow NO_2^- \left(adsorbed\right) \tag{4.8}$$

$$NO_2^- \left(adsorbed\right) + O^- \left(adsorbed\right) + 2e^- \rightarrow NO\left(gas\right) + 2O^{2-} \left(adsorbed\right) \tag{4.9}$$

It has been observed that under the gas exposure, response of the POS sensors is only the square root of the NOS sensors' response with identical morphology and environment [23]. Although the POS gas sensors have less sensitivity than the NOS gas sensor, they have the advantages of being less humidity dependent, high catalytic reactive, and so on [16]. Thus, the selection of appropriate material for sensor preparation depends on application-specific environments.

In some cases, the target analyte gas molecules react with active material instead of an oxygen adsorption mechanism. When SnO_2 is exposed to H_2S, H_2S is adsorbed into SnO_2 following the direct chemical reaction between SnO_2 and H_2S.

$$SnO_2 + 2H_2S \rightarrow SnS_2 + 2H_2O \tag{4.10}$$

As SnS_2 has a narrower bandgap compared to SnO2, the adsorption of H2S leads to a decrease in resistivity. However, the time-consuming reaction process between SnO_2 and H_2S results in reduced speed of the device [1]. This chemisorption results in conductivity modulation due to the chemical change in the active material. Apart from these two categories, there is a physical adsorption/desorption process in which the gas molecules are adsorbed into the material without any chemical rearrangement.

4.2.1 Gas Diffusion in the Active Material

The Knudsen diffusion and Arrhenius equation, the diffusivity and surface chemical reaction rate are defined as [24,25]:

$$D_{KA} = \frac{d}{3}\sqrt{\frac{8RT}{\pi M_A}} \tag{4.11}$$

$$k = A \times e^{-\frac{E_a}{RT}} \tag{4.12}$$

where d = pore diameter of the nanostructure, R = gas constant, T = absolute temperature, M_A = molecular weight, A = pre-exponential factor, and E_a = activation energy.

At low temperatures, the diffusivity is very small, and gas is primarily adsorbed on the outer surface, and surface chemical reaction becomes the dominant factor. However, with increased temperature, the gas diffusivity is gradually enhanced and becomes the primary parameter to control the sensitivity [1].

4.3 SMO GAS SENSOR DESIGN

Figure 4.4 displays the diagrammatic representation of an experimental gas sensing setup. Several functional units work together to measure the target gas's presence and concentration in a typical gas sensing system. Two gas cylinders contain the target gas and inert carrier gas (N_2) separately. The

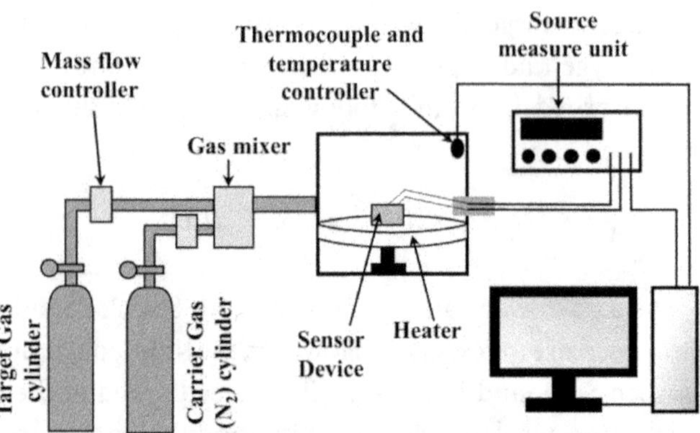

FIGURE 4.4 Schematic of an experimental gas sensing setup.

concentration and flow of these gases are regulated by a mass flow controller and gas mixer unit. In the sensing chamber, the sensor is associated with a heater unit. The thermocouple and temperature controller unit monitor the chamber unit as per requirement. The probe tips from the probe station complete the connection between the device and the source measure unit (SMU) and thus PC. The variation in device resistance may be recorded using SMU with alternating gas concentration temperature and chamber temperature. Once the resistance value becomes steady, the device recovery and the recovery characteristic are observed by the opening chamber.

4.3.1 Parameters for Evaluating the FoMs

Several parameters are used for determining the commercialization aspect of a gas sensor. The most critical parameters are sensitivity, selectivity, and stability. These three parameters are the primary performance evaluating criterion whose comparisons are carried out to select a superior gas sensor. However, some more parameters for critical examination are response and recovery time, working temperature, the limit of detection, dynamic range, and so on [19].

Sensitivity: It is described as the modulation in resistance due to exposure in the target gas analyte. It is expressed by R_g/R_a or R_a/R_g where R_g and R_a represent the resistances of the gas sensor in the gas environment and air, respectively. The first parameter (R_g/R_a) is used when an oxidizing (reducing) gas is introduced to NOS (POS). In both cases, we observe that under gas exposure, the R_g parameter is enhanced. Similarly, when a reducing (oxidizing) gas is introduced to

NOS (POS), R_a/R_g is used to measure responsivity. The sensitivity is described as the modulation in resistance per unit variation in analyte gas concentration.

Selectivity: The sensor's selectivity is defined as its ability to detect a specific gas among various interfering gas molecules' ensemble environments. This parameter is analogous to the signal-to-noise ratio of a communication system. This parameter becomes very important when a gas sensor is employed in a hazardous gaseous environment where different toxic gases can significantly harm the living organism. This parameter is expressed as the ratio of signal of target gas and an interfering signal.

$$S = \frac{S_t}{S_i} \tag{4.13}$$

Here, S_t is the sensor signal of target gas, and S_i is the sensor signal of interference gas. As the sensing principle is almost the same for most oxide semiconductors, the gas sensors suffer from low selectivity.

Stability: Stability represents the reproducibility of the sensor response under a varying environment. Two categories of stability are associated with a gas sensor. When a gas is introduced to a sensor, its response should be reproducible for a defined period during its working environment under high temperature. This is known as active stability. The device also needs to retain its sensing attributes during its whole lifetime, even in ambient environment/non-active modes. The entire lifetime of a sensor can be ~17000–26000 h of operation [18, 26].

Response and recovery time: Response time is described as the time to attain 90% of the final resistance variation under exposure to target gas. Recovery time indicates the time taken to return to 10% of the initial resistance value when the gas environment is removed. Low values of response and recovery time are required for faster operations.

Working temperature: The gas sensor response is very much dependent on its environment temperature as the diffusivity of a gas molecule is closely related to temperature. At operating temperature, the sensor exhibits maximum sensitivity.

Limit of detection: The minimum amount of a gas detected by a sensor at a given temperature is a known limit of detection.

Dynamic range: Dynamic range is the scale of the analyte gas concentration between the lowest and the highest detectable gas concentrations [27].

4.3.2 Factors Affecting Gas Response and Sensitivity

The applicability of the semiconducting metal oxide (SMO) gas sensors is based on the sensitivity of these semiconductor's electrical conductivity to the surrounding gaseous analytes present in the working atmosphere, which is dependent upon the charge transfer between SMOs and reactive gaseous species such as NOx, CO, H_2S, and NH_3 [28].

By changing the sensors' microstructure, mainly grain size and porosity, sensitivity can be controlled [29,30]. Xu et al. explained the influence of grain size on SMO gas sensor sensitivity [31]. According to the proposed model, neighbouring crystallite contained in the sensors are connected by necks. The grain boundaries (GB) separate the neighbouring crystalline grain regions with different planes and directions. If grain size = D and depletion layer's width = L, for D>>2L, the maximum volume remains neutral during gaseous analyte–surface interaction [Figure 4.5(a)]. Thus, charge transfer is mainly governed by GB barriers because the charge transfer from one to another grain is carried out via the GB barrier. Thus, GB control the influence of gaseous analyte on the conductivity of the SMO sensor. Hence electrical resistivity varies exponentially with barrier height (qV_b).

$$\sigma \propto e^{\left(-q|V_B|/kT\right)} \tag{4.14}$$

Here V_B term stands for grain boundary potential, q is the electron charge, and k and T are Boltzmann constant and temperature, respectively. Further, the relationship between qV_B (barrier height) and charge density trapped at the crystalline surface is given by

$$q|V_B| \propto \left(N_t^-\right)^2 \circ \tag{4.15}$$

Here N_t^- is trapped charge density. Trapped charge density depends upon the charge-transfer interaction with gaseous analyte such as NO_x, CO, etc. [32,33].

Thus, the conductivity will be affected by barrier height and trapped charge density, depending on the gas composition. Hence, for large grain size (D >> 2L), GB barriers control the mechanism of gas sensing. Moreover, GB barriers are not grain size (D) dependent. Therefore, sensitivity is independent of D for its large magnitude. With a smaller D, the depth of the depletion region into grain increases. The core region, which is more conducting, becomes less wide. When D approaches 2L but still more than 2L (D>> 2L), the depletion region with low conductivity surrounds every neck resulting in narrow conducting channel, as shown in Figure 4.5(b). Due to these constricted conductive channels' formation, conductivity depends on the GB barriers and cross-sectional area of these channels. This cross-sectional area is proportional to $(X\text{-}L)^2$, where X denotes neck diameter. Since L is proportional to N_t^-, N_t can be controlled by the interaction of the gaseous phase with the surface. Therefore, the effective cross-sectional area depends on the atmospheric gas composition due to its dependency on the L. Thus, the current constriction effect combined with the GB barrier effect influences the sensitivity of the sensor. Thus, the sensitivity of the sensor increases compared to the earlier case of D >> 2L. In addition to this, in the present case, gas sensors' sensitivity to the gas becomes dependent on grain size D. Thus, when grain size D approaches 2L, the neck becomes the dominating factor affecting the sensor's sensitivity [31]. Therefore, in this case, sensitivity is neck-controlled.

4.3.2.1 Case III: D < 2L

In this case, the crystallite gets completely depleted of mobile charge carriers, as shown in Figure 4.5(c). Due to the depletion region's extension throughout the crystallite, conducting channels vanish, and the energy bands become almost flat throughout the neighbouring connecting grains [34]. Thus, the energy barrier deciding the conductivity now disappears, and charge transport is therefore solely grain controlled. The highest sensitivity is observed in this case. However, when the grain size is smaller than the 2L, the steep increase in the sensitivity with a decrease in grain size still needs to be explored for clear understanding.

Based on the dependency of sensitivity on grain size explained above, it was concluded that the smaller grain size is favourable for enhancing sensitivity [31]. However, excessive decrement in grain size may result in structural instability. Another way to achieve enhanced sensitivity is by tuning the porosity and microstructure of the SMO gas sensors. Metal oxide with high porosity and large surface area possesses high sensitivity. The charge carrier concentration of a sensing material can be tuned at any operational

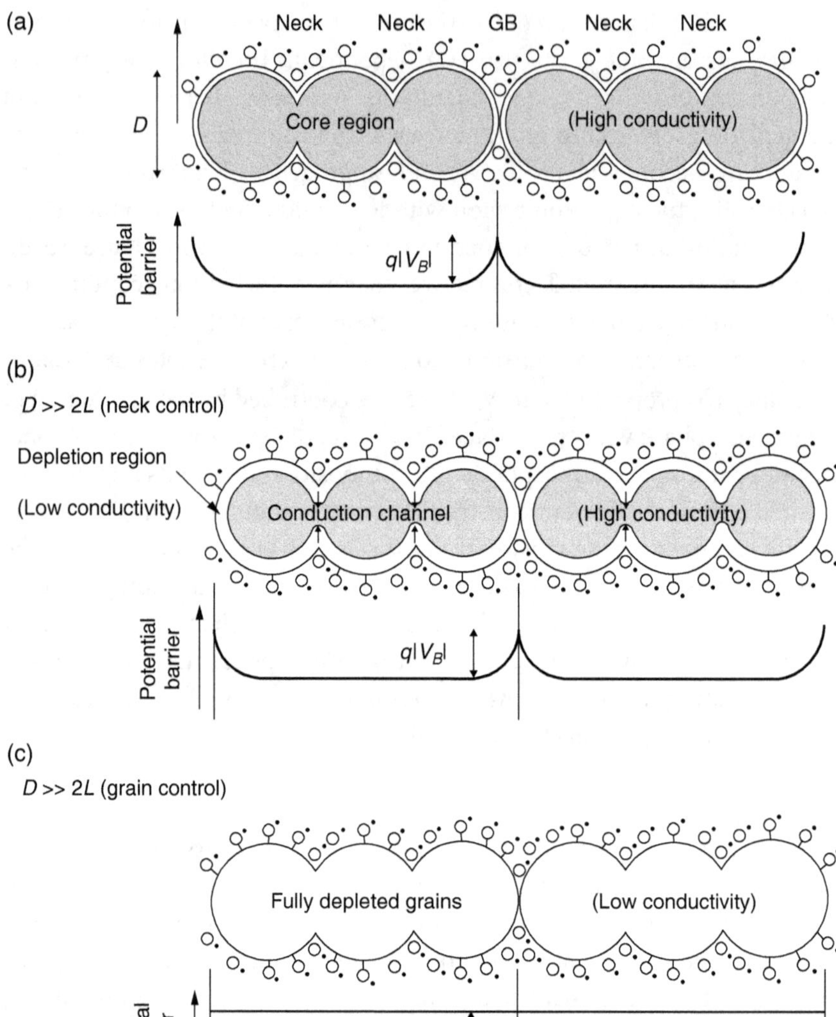

FIGURE 4.5 Schematic model of the effect of the crystallite size on the sensitivity of metal-oxide gas sensors: (a) D>>2L (GB control); (b) D≥2L (neck control); (c) D<2L (grain control) [Reprinted from [33] Effect of grain size on the sensitivity of nanocrystalline metal-oxide gas sensors. Avner Rothschild, and Yigal Komem, *Journal of Applied Physics*, 95, 6374 (2004) with permission from American Institute of Physics.]

temperature to get a desirable high response to exposed gases. Since it is quite difficult to control non-stoichiometry under ambient conditions at any fixed sensing temperature, SMO sensors' high sensitivity can be achieved by introducing dopants or any impurity. Based on the dependency of sensitivity on grain size explained above, it was concluded that the smaller grain size is favourable for enhanced sensitivity. However, decreasing "D" excessively might result in structural instability. Moreover, changing the porosity and microstructures of SMO gas sensors can also improve the sensitivity. SMO with high porosity and large surface area possesses high sensitivity [35, 36]. In addition to this, the SMO sensor's sensitivity can be changed by introducing dopants or any impurity. The mechanism responsible for the change in the sensitivity by adding some impurities is not yet fully understood. Charge carrier concentration of a sensing material can be optimized or tuned at any operation temperature to get a desirable high response to exposed gas (more sensitivity). Since it is impossible to control the oxygen non-stoichiometry at a fixed sensing temperature in an ambient, doping might be proven useful for improving sensitivity [37, 38].

4.3.3 Factors Affecting the Stability

Stability is a crucial issue in SMO sensors that needs to be addressed, specifically. Sensor stability is distinguished in two ways: one of them is defined as maintaining reproducibility of their characteristics under working conditions such as analyte exposure and high working temperature. This type of stability is termed active stability. Another type of stability includes retaining its characteristics such as selectivity and sensitivity for a period under a normal condition such as ambient humidity and room temperature; this is known as conservative stability. The stability of the sensor in the case of the metal oxide sensing material is low, leading to signal drift that sometimes may demand frequent recalibration and replacement of the sensor. Moreover, it might result in false alarms due to the uncertainty of the results. Therefore, the stability of these gas sensors is a matter of concern. Nanostructured metal oxide having small grains such as rods and nanotubes generally undergo degradation due to their very high reactivity. However, the issue of the instability of the sensing material cannot be solved by a unified approach, but it could be done up to some extent by post-process treatment such as annealing and calcination and by following strategies for reduced working temperature. It is also reported that doping with metal particles or carbon nanotubes and the use of mixed oxides can be proven useful to attain the increment in the stability of the sensing material [39, 40].

4.3.4 Response Time and Recovery Time and Factors Affecting Them

Short recovery time and rapid response are desirable for the real-time application of sensors due to the explosive character and high toxicity of some target gases. Response time is described as the time to attain 90% of the final resistance variation under exposure to target gas. Recovery time indicates the time taken to return to 10% of the initial resistance value when the gas environment is removed [41]. However, in few of the reports, 63% of the stable change in resistance is defined as the response time [42]. Porous structures can facilitate the abundant diffusion of the gas species thus reducing this delay.

4.3.5 Selectivity (Cross-Sensitivity) and Factors Affecting Them

The active material can be chosen such that it has very high selectivity to one of the analytes and relatively very low or nearly zero to other expected analytes present in the mixture. Another approach to achieve high selectivity is to distinguish between the properties of the different gases, which could be done either using the array of sensors according to the expected gases present in the working atmosphere or by temperature modulation sensor for different gases. The addition of some external impurities or synthesizing hybrid SMO can increase the gas sensor's selectivity due to the involvement of different materials with selectivity to different gaseous analytes. The sensor sometimes exhibits the cross-sensitivity being sensitive to multiple gases limiting their practical application. The general strategies to increase the selectivity include tuning the sensing temperature, using additives and heterostructures, functionalizing with noble metals, and using filters [43]. Metal-organic frameworks (MOFs) can enhance the selectivity due to their functional pores [44]. Due to adjustable pore size, MOFs are capable of the selective separation of gases [45, 46].

4.3.5.1 Sensing Temperature

The temperature-dependent properties such as reaction and diffusion of the target gases to be sensed affect gas-sensing performance; therefore, the temperature influences the gas-sensing properties. The reaction rate of the target gas is the deciding factor for the response at lower temperatures; however, at high temperature, the gas molecules' diffusion limits the sensor's response. At intermediate temperature, both factors (diffusion and rate of chemical reaction) contribute resulting in a maximum response. In general, the operating temperature for gas sensing in metal oxide lies in the temperature range of 25–500°C [17, 47]. In this range, the temperature variation affects the formation of chemisorbed species. The equations related to the chemisorbed oxygen ion species (O^-, O_2^-, O^{2-}) are as follows:

$$O_{2(gas)} \rightarrow O_{2(ads)} \tag{4.16}$$

$$O_{2(gas)} + e^-_{surface} \leftrightarrow O^-_{2(ads)} \left(< 100°C\right) \tag{4.17}$$

$$O_{2(ads)} + 2e^-_{(surface)} \leftrightarrow 2O^-_{(ads)} \quad \left(100 - 300°C\right) \tag{4.18}$$

$$O^-_{(ads)} + e^-_{(surface)} \leftrightarrow O^{2-}_{(ads)} \left(> 300°C\right) \tag{4.19}$$

Depending on the chemical composition, target gas, and morphology, each sensor exhibits an optimum sensing temperature [48]. For some specific gas-sensing applications, high-temperature gas sensing is required [49]. However, low-temperature sensing is favourable to serve the purpose of low energy consumption and the applicability of the gas sensors in remote areas.

4.3.5.2 Limit of Detection (LOD)

The limit of detection can be expressed as follows:

$$LOD = C_{min} = f^{-1}\left(R_{min} \right), \quad R_{min} = R_0 + 3\sigma_0 \tag{4.20}$$

where $f^{-1}(R)$ stands for calibration function's inverse. C_{min} is the minimum detectable concentration. The R_{min} (minimum sensor response) corresponds to the noise. LOD can be calculated as "3 × $noise_{rms}$/slope", where $noise_{rms}$ is termed as the standard deviation of the signal and slope is response vs. gas concentration curve's first derivative [50, 51].

4.4 MEASUREMENT SYSTEM

Nanowires-based gas sensors are of great importance for their high selectivity, ultra sensitivity, less power consumption, and less response time. Due to the high SVR of the nanowires, a very low concentration of the gas is sufficient to alter the sensing material's electronic properties, resulting in superior sensitivity and ultrafast response. Due to lightweight, less power consumption, and miniaturization, this is well suited for a wide range of gas detection. Nanowires-based gas sensors are sensitive to very less amount of gas, which is useful for their operation at low temperature.

4.4.1 Nanowire Synthesis and Nanofabrication

There are mainly two approaches for synthesizing nanowires: top-down and bottom-up approaches to attain dimensionally homogeneous well-ordered nanostructures [52]. Furthermore, in the top-down technique, while generating integrated circuits, parts are produced in place and assembled, so additional assembly is not required. However, this approach has some shortcomings, as these approaches work well at the microscale but implementation becomes increasingly difficult at nanoscale. Though patterning techniques such as lithography are widely used for manufacturing, these techniques suffer the disadvantages of high cost, specifically for fabricating structures with high resolution. In contrast, in the bottom-up approach, individual atoms or molecules agglomerate to fabricate the desired morphology. Numerous fabrication processes are associated with the bottom-up approach, including evaporation, vapour-liquid-solid growth, liquid phase epitaxy (LPE), chemical vapour deposition (CVD), pyrolysis, layer-by-layer assembly, condensation (atomic or molecular), and so on. The bottom-up approaches can be used conventionally for the fabrication of nanoscale structures in a cost-effective way. A transition to an inactive state might occur after the interaction of metal oxides with the gas analyte with a large impact on sensors' conductance. In the case of nano-dimensional grains, the surface regions trap most charge carriers. Therefore, the conduction is governed by a few thermally activated carriers. Consequently, it is challenging to fabricate small-sized materials that can retain their stability at high temperatures even after long-term operation. In polycrystalline sensors, high temperature is needed for surface reaction, which results in grain growth by coalescence, thus restricting the attainment of stable materials.

Metal oxide nanostructures are more useful relative to the thin film due to their high SVR, more stability due to high crystallinity, and more possibility of modulation of operation temperature to attain selectivity. Semiconductor nanowires became a field of great interest within the nano-sensing research community. Several experimental methods are used for the synthesis of the SMO nanowires. The bottom-up approach's advantage includes high purity of the material, low-cost equipment, and easy control for doping and junction formation. But there are a few disadvantages of this technique, too, such as integration on the substrate becomes difficult. A combination of bottom-up and top-down approaches is the best solution.

4.4.1.1 Synthesis of Metal Oxide Nanowires

The use of metal catalyst growth mechanism in SMO nanostructures can be classified into two classes: VLS (vapour-liquid-solid) growth and VS (vapour solid) growth. In VS growth, gold and nickel nanostructures are introduced as the catalyst [53–55]. The supersaturated atoms segregate as the metal oxide nanostructures near the melting point of the metal catalyst. Relative to the VS growth, VLS mode has the benefits of (i) a wide range of oxide material for selection and (ii) simplicity of the process. The shortcomings of the VLS growth mode are the possibility of contamination of the catalyst by reacting with the target during high-temperature growth [56, 57]. Therefore, it tried to utilize the metal component corresponding to the target metal oxide as the self-catalyst [58–61]. This was attempted by the decomposition of metal oxide to achieve a high-purity product without contamination. However, direct decomposing of the metal oxide requires a very high temperature, which is too tough to control practically. Generally, heat furnaces suffer the limitation of the attaining temperature restricting their application for the metal oxide with a decomposition temperature of more than 1500°C. To solve this issue, the carbon-thermal decomposition method is used. In this method, the decomposition temperature can be lowered by mixing graphite powder with the SMO. However, this method also involves contamination due to the formation of metal carbide [60, 61]. In some cases, electron beam evaporation was used to decompose high melting point materials [62–64]. To decrease the growth temperature, low melting point catalysts such as indium and bismuth were used [65, 66]. However, in the case of SMO, heterogeneous catalysts are not preferred. Usually, metal corresponding to SMO can be decomposed during growth. Comparative to other conventional techniques and tools used in the development of nanostructures, the vapour pressure needed for self-catalysis can be easily achieved by an electron beam [67]. In conclusion, electron beam deposition is superior to other tools used for the nanowire growth by using self-catalysis.

4.4.1.2 VLS Growth

Wagner and Ellis first introduced this growth method. They grow Si whiskers by using Au droplets as the catalyst [53, 68]. This method is inexpensive and simple relative to other methods for growing nanowires with a high aspect ratio [69]. In general, the metal catalyst is used for the crystallization [70]. At some fixed temperature, growth time decides the length of the nanowires. Moreover, the metal catalyst droplet dimension decides the radius of the nanowires. In this method, during growth, firstly the metal catalyst is melted into a liquid alloy droplet. The nanowire material vapours prefer to get absorbed at the liquid droplet sites; the precipitation process at the solid–liquid interface results in the deposition of nanowires. At supersaturation of alloy droplets, precipitation of the source metal starts taking place, and under oxygen flow, metal oxides grow with orientation resulting in the 1D nanowires [70]. Ideally, a eutectic alloy is expected to be formed by the catalyst. The growth temperature is kept in a range between the eutectic point and the material's melting point. Both physical and chemical methods can be utilized for the generation of vapours required for the growth of the nanowires [71].

4.4.1.3 Growth of Branched Nanowires

These nanostructures are also named nano trees or nanoforests. These nanowires have a three-dimensional (3D) structure with many heterojunctions or homojunctions and electron transport pathways. For gas sensing and detection at the sub-ppm level, branched 3D nanowires are preferable candidates to 1D NWs and nanoparticles due to their superior SVR and structural hierarchy [72, 73]. Moreover, branched nanowires provide an enhanced conduction path through the branches [74]. The typical methods used to fabricate branched nanowires include self-catalytic growth, VLS growth, and VLS combined with screw dislocation [75]. Among these methods used, VLS is the most popular. It involves the following sequential steps: firstly, deposition of the nanowires followed by metal catalyst onto the as-synthesized primary nanowires. Then another growth of branched nanowires takes place. The branched nanowires' density depends on the primary nanowires' catalytic amount, while the length is related to the growth time [72]. Thus, VLS growth is preferred to synthesize branched nanowires with desirable features and crystalline quality. However, this method and much more vapour phase growth method demand a high temperature [72].

4.4.1.4 Synthesis of C-S Nanowires

A C-S nanowire consists of a nanowire core surrounded by a shell made of a different material. It is one of the classes of composite material. In this method, a shell of some other material is coated around the core of the nanowire. The characteristics of the C-S nanowires can vary from the properties and shell material. The ideal synthesis method should have the versatility of highly crystalline shell deposition and control over the thickness of the core material. Moreover, it should be cost-effective, safe, and with minimum detrimental effect [76]. In the case of synthesis of C-S nanowires, the bottom-up approach is much more beneficial than the top-bottom approach to field significantly fine structures and energy loss minimization [77]. Among the available fabrication techniques (such as laser-induced assembly, CVD, etc.), atomic layer deposition (ALD) provides good control of shell thickness. This technique is based upon vapour phase technology and can be used to synthesize nitrides and SMOs with a control on thickness even at the sub-nanometre level [78–81]. The main advantage of ALD is uniformity and fine conformal control over thickness, making this technique advantageous for the 1D NWs coating [82–84]. These coatings, deposited using ALD, are continuous, smooth, pinhole-free, and conformally coat the primary core. Schematic presentation of the technique used for the synthesis of nanowires growth is shown in Figure 4.6.

4.4.2 Single Nanowire Gas Sensor

The development of single NW-based gas sensors requires complex fabrication methods and very careful observation. Nanowire's diameter, the synthesis procedure, and the reaction occurring on the nanowire's surface can greatly affect the performance of the gas sensor [85]. There are several fabrication issues in realizing these gas sensors, such as electrical contact formation. Due to the nanometre resolution of the e-beam lithography, this technique is useful for forming electrical contacts. Due to the complexity of single NW gas sensors' fabrication process, these sensors' commercialization is challenging [86]. Tonezzer proposed SnO_2 NWs-based gas sensors fabricated via CVD [85]. NWs with varying diameters were dispersed on the substrate, and NO_2 sensing studies were performed. The expression used for determining the relationship between the depleted zone's depth and diameter of the nanowire is given by [85]:

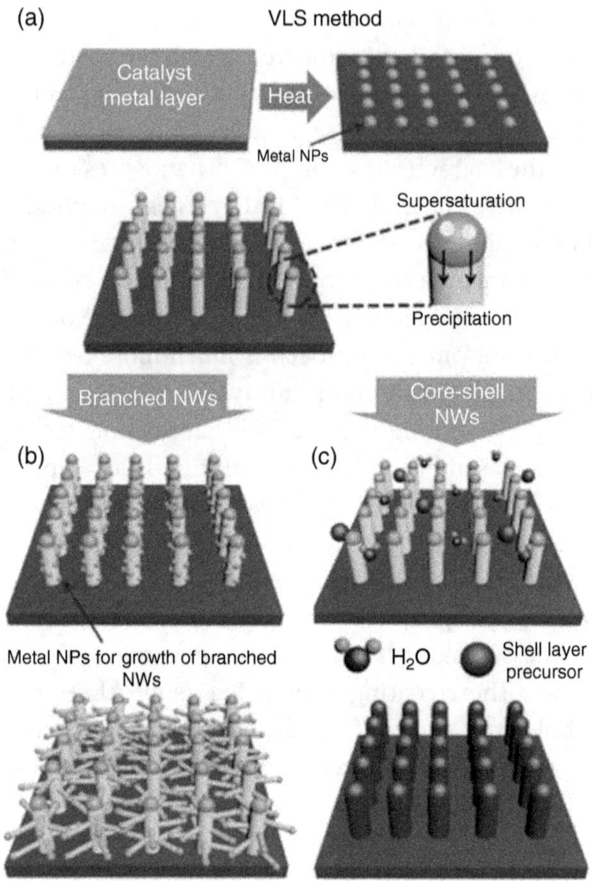

FIGURE 4.6 Schematics of different techniques for the growth of NWs: (a) VLS method; (b) branched-NW growth; and (c) C-S NW growth. [Reprinted from [19] Resistive gas sensors based on metal-oxide nanowires. Ali Mirzaei, Jae-Hyoung Lee, Sanjit Manohar Majhi, Matthieu Weber, Mikhael Bechelany, Hyoun Woo Kim, and Sang Sub Kim. *Journal of Applied Physics*, 126(24), (2019) with permission from American Institute of Physics.]

$$Response = Constant \times \left(\frac{R}{R - L_D} \right)^2 \tag{4.21}$$

It was found that NW with a small diameter yields a comparatively higher response for NO_2. The influence of nanowire radius on the sensor's performance was confirmed by Lupan et al. [87]. It was reported that the ZnO nanowires with 100 nm diameter exhibited an enhanced sensitivity.

This enhancement might be due to the Debye length comparable to the diameter, resulting in resistance modulation [87]. Generally, networked and multiple nanowire-based gas sensors have better performance than a single nanowire-based gas sensor [63]. This might be attributed to the homojunction in numerous and networked nanowires, which results in strong resistance modulation. Zhang et al. proposed In_2O_3-based gas sensor with ppb level response for the NO_2 gas at a temperature of about 25°C [88]. However, the multiple nanowire gas sensor's responses were higher as well as reliable. Moreover, the fabrication of multiple nanowires-based gas sensors was quite simple [89]. Platinum-coated ZnO-based single nanowire gas sensors showed high sensitivity and selectivity at room temperature hydrogen sensing. Thus, metal-coated single nanowire gas sensors offer better performance compared to the unmodified ones [90].

4.4.3 Branched Nanowire Gas Sensor

Bang et al. reported high-performance Bi_2O_3 branched SnO_2 nanowires fabricated by the VLS method for sensing NO_2 gas [91]. It was highly responsive for NO_2 gas of concentration ranging from 56.92 to 2 ppm. This high performance may be due to the increased SVR resulting from the branching of the structures. Kim et al. observed an increased response of CuO-modified branched SnO_2 nanowires-based gas sensor when exposed to the hydrogen sulphide gas [92]. CuO is a semiconducting state, while CuS is a metallic state. Thus, conversion from the semiconducting to the metallic phase was found to be the reason responsible for the superior performance of the sensor due to the modulation of resistance significantly. In contrast, the same sensor was less responsive to NO_2 gas [92]. Kim et al. proposed Co NPs tipped on the SnO_2 nanowires with the branching of ZnO nanowires using the VLS method [Figure 4.7(a), 4.7(b)] [93]. The influence of Co embellishment on the branched nanowire gas sensor's sensing mechanism is schematically shown in Figure 4.7(c) and 4.7(d). The analyte gas (NO_2) migrates towards the ZnO surface after absorption on Co nanoparticles resulting in an increase in the resistivity. Highly porous ZnO nanowires grown by the VLS method were subjected to the cation exchange reaction in $CoCl_2$, leading to the transformation of ZnO nanowires into the CoO nanowires. The Co-doped ZnO nanowire-branched structure was then obtained by co-deposition of branched ZnO nanowires, as shown in Figure 4.7 [94]. Thus, due to the number of homojunctions present in

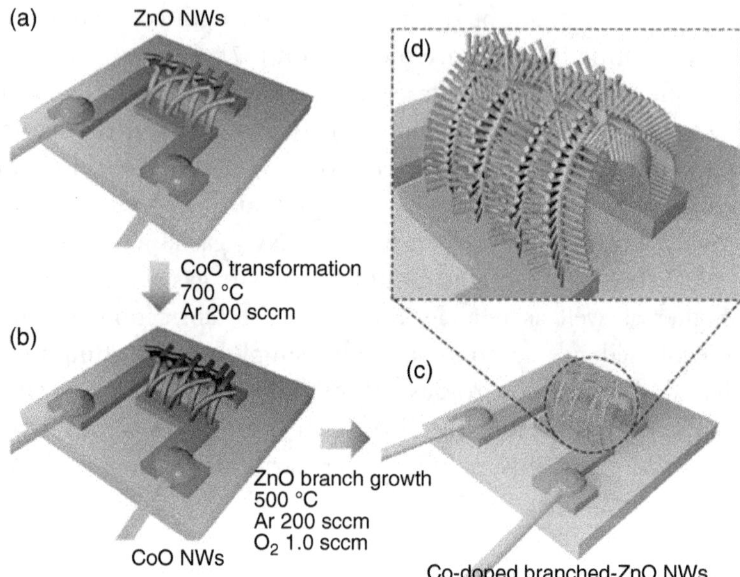

FIGURE 4.7 (a) ZnO NWs; (b) transformation of ZnO NWs into CoO NWs; and (c) and (d) Co-doped ZnO NW branches. [Reprinted from [94] Co-doped branched ZnO nanowires for ultraselective and sensitive detection of xylene. Hyung-Sik Woo, Chang-Hoon Kwak, Jae-Ho Chung, and Jong-Heun Lee. ACS applied, materials & interfaces, 6(24):22553–22560, (2014) with permission from American Chemical Society.]

ZnO nanowire branches, a higher initial resistance results relative to the pristine ZnO nanowires. It was found that in the presence of p-xylene, resistance modulation takes place significantly, resulting in the sensor's high responsivity. Moreover, the catalytic activity of the Co dissociates p-xylene into the small gases, which have high reactivity compared to the p-xylene. The gas sensor's response was improved when less reactive Xylene decomposed into highly reactive gases. Further in a similar study, Ni-doped ZnO nanowires-based sensor exhibited a very high sensitivity for a 5-ppm concentration of p-xylene at a temperature of about 400°C [95]. This gas sensor was significantly selective to the p-xylene in comparison to other gases. A high sensitivity of the branched nanowires compared to the unbranched ZnO nanowires-based gas sensor was reported. The porous 3D structure and large SVR due to branching might be the reason for the high response.

4.5 ENHANCEMENT OF PROPERTIES OF SMO GAS SENSOR

There are various strategies to enhance the properties of the gas sensor.

4.5.1 Change in Microstructure and Morphology

As discussed earlier, microstructure and morphology are factors that significantly affect the gas sensor's sensitivity. Sensitivity increases with a decrease in the grain size [96]. Sensitivity drastically increases when the grain size becomes smaller than the space charge depth. The morphology is also one of the deciding factors in determining the response of the gas sensor. Branched structures are more favourable as compared to the unbranched ones to realize the highly responsive gas sensor. Highly porous structures with a high surface area are desired to enhance the gas sensors' sensing ability. A variety of SMO nanostructures such as SnO_2, WO_3, In_2O_3, NiO, and TiO_2 were used for the sensing application [97–102]. Recently, efforts were made to fabricate several morphologies to improve sensing properties. Based on the morphology, semiconducting oxide materials can be classified into 0, 1, 2, and 3-dimensional nanostructures [103–105]. The properties of the SMO change with the dimension; therefore, the relationship between dimensionality and sensing properties is as follows

4.5.1.1 Zero Dimensional

These materials offer more surface area for absorption of the analyte gas and thus provide an opportunity to enhance the gas sensing properties [106–110]. Moreover, nanoparticle agglomeration and surface sensitivity affect sensing properties. The gas absorption by active sites at the sensing materials' surface results in the reaction of gas. These reactions occur through the process of adsorption, desorption, and charge transfer. The adsorption sites at the surface of sensors affect the performance of the gas response, which might be due to an increase in oxygen adsorption due to increased surface-active sites. SVR greatly influences the concentration of active sites; this boosts more target gas molecules to participate in the oxidation and reduction reaction. Wng et al. reported superior benzene and nitrogen dioxide sensing performance by octahedral nanoparticles of Cu_2O compared to Cu_2O nanorods [111]. The availability of more active sites due to the large SVR was attributed as the main factor leading to the high response. Quantum dots were frequently used for gas sensing applications due to their high SVR and control over the nanoparticle

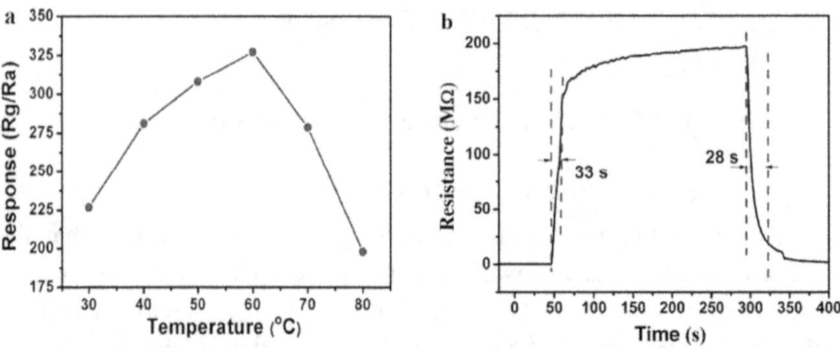

FIGURE 4.8 (a) Response of the sensor based on In_2O_3 nanoparticles to 5 ppm NO_2 as a function of operating temperature. (b) The typical dynamic response of the sensor to 20 ppm NO_2. [Reprinted from [113] Facile synthesis of In2O3 nanoparticles for sensing properties at low detection temperature. Bingxin Xiao, Fei Wang, Chengbo Zhai, Pan Wang, Chuanhai Xiao, and Mingzhe Zhang. Sensors and Actuators B: Chemical, 235:251–257, (2016), with permission from Elsevier.]

size. A highly responsive and energy-efficient WO_3 colloidal quantum dots-based gas sensor was reported for H_2S gas sensing; this sensor's operating temperature was very low (80°C). This high response might be attributed to the miniaturization of crystal dimensions resulting in a higher SVR [112].

It is observed that the coalescence of the nanoparticles affects gas sensing performance. The particle dimensions need to be uniformly distributed. Xiao et al. reported a gas sensor based on In_2O_3 nanoparticles of 20–30 nm diameter with a uniform size distribution of nanoparticles which showed its peak response at 60°C [Figure 4.8(a), 4.8(b] [113]. The uniformly dispersed nanoparticles were able to sense the analyte gas quickly.

This sensor has the potential to be used for NO_2 sensing. Xu et al. proposed a model according to which any sensor's response remarkably increases when $D < 2L$. The size effect on the sensors' response is already discussed in the section sensitivity and factor. Liang et al. prepared α-Fe_2O_3 nanoparticles by using the microemulsion method; the nanoparticles' reported size was about 3 nm. The value of L was 44 nm in this case and fulfilled the criteria of $D \leq 2L$. The depletion of the grains resulted in change of conductivity, and hence sensing performance was improved. This shows the better sensing capability of 0D nanoparticles.

4.5.1.2 1D Sensing Material

1D nanomaterials are widely used in design of gas sensors due to the high aspect ratio. The length of these materials can be in the millimetre range; however, the width is limited to the 1 to 100 nm range [114]. Various 1D nanostructures such as nanowires, nanobelts, nanotubes, and so on were explored by the researchers [115–126]. Due to their high SVR, nanostructures with a high aspect ratio have superior sensing abilities. Lupan et al. showed the effect of the diameter of ZnO nanorods on hydrogen detection [127]. The ZnO nanorods with small diameter were found to have more suitable properties. Besides this, intercrossing of nanowire and junctions provides a conducting path to enhance the potential barrier height resulting in a change of resistance. Nanotubes exhibit porosity as well as large surface area as they have hollow structures. It was evident from the literature that for ultra-long 1D nanostructured materials, the electron transport takes place along an axial direction resulting in superior gas sensing characteristics. Among 1D nanomaterials, nanofibres were found to be of great interest. Nanofibres possess a high SVR due to the high aspect ratio (small diameter and long length). It was shown by Yoon et al. that a decrease in interparticle connectivity reduces the gas sensing response [128], which was due to both good interparticle connectivity and high surface area; thus, the increased surface area is as vital as particle connectivity.

4.5.1.3 2D Sensing Materials

Due to their unique morphologies, 2D nanostructures can be used in lithium-ion energy storage devices [129], photo electrochemistry [130], gas sensors [131], and photocatalyst [132]. Compared to dense nanostructures, porous nanostructures facilitate gas interaction with the sensor surface and penetration of gas into sensing materials. Porous nanostructured materials can be classified based on pore size [133]. If the pore's size is smaller than 2 nm, it is called micropore; however, if the pore size is more than 50 nm, it is microporous. The materials with pore sizes between 2 and 50 nm are known as mesoporous materials. Mesoporous nanostructures combined with 2D nanosheet improve the gas sensing performances. Mesoporous In_2O_3 ultrathin nanosheet synthesized by Wang et al. showed enhanced response with a very fast response time of 4s at low operational temperature with a 10 ppb LoD [134]. The ultrathin nanosheet of thickness 3.7 nm resulted in more active sites to improve sensing performance. Moreover,

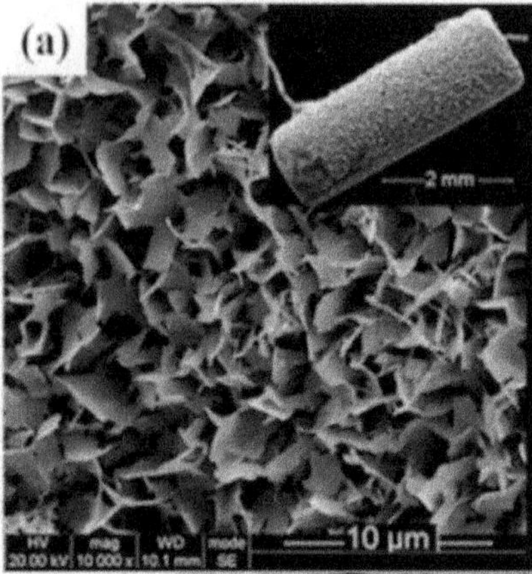

FIGURE 4.9 (a) SEM image of samples with magnification of 10,000 times (inset indicates the panoramic image of the CO_3O_4 on the surface of alumina tube), (b) SEM images with magnification of 40,000 times. [Reprinted from [135] High precision NH_3 sensing using network nanosheet CO_3O_4 arrays-based sensor at room temperature. Zhijie Li, Zhijie Lin, Ningning Wang, Junqiang Wang, Wei Liu, Kai Sun, Yong Qing Fu, and Zhiguo Wang. Sensors and Actuators B: Chemical, 235:222–231, (2016), with permission from Elsevier.]

due to the mesoporous structure, the path gets shortened leading to more gas diffusion. The morphologies of crossed nanosheets of Co_3O_4 are shown in Figure 4.9(a), (b) [135]. Due to the large surface to volume ratio, crossed nanosheets offer electrical channels; thus, networked structure exhibits good sensing abilities. It was observed that Co_3O_4 nanosheet arrays exhibit good response with high sensitivity to a low concentration of ammonia. Sensitivity to low concentration might be attributed to the intersection between open structure nanosheets. This type of morphology prevents aggregation with large SVR to facilitate the reaction between the target gas and sensing surface. Thin films also fall into the 2D material category, and their surface roughness greatly enhances the performance of gas sensors. It was also reported that CVD grown ZnO layer showed efficient CO sensing at 300°C [136]. Thus, the morphology, surface roughness, significantly affects the gas response. The effects of 2D nanostructures' morphology on the gas sensing performance are summarized Table 4.2.

4.5.1.4 3D Materials

3D nanostructures are of great significance to detect toxic gases due to their high SVR, an abundance of adsorption sites, and swift movement of carriers

TABLE 4.2 Summary of the Gas Sensing Performance of 2D Nanostructures with Various Morphologies to Different Gases

Morphology	Gases	Materials	Conc. (pp)	LOD(ppm)	Temp(°C)	t_{res}(s)	T_{rec}(s)	Resp.	Ref.
Nanosheet	Acetone	Co_3O_4	100	1.8	150	NA	NA	11.4^b	137
Nanosheet arrays	NO_2	ZnO	10	NA	180	3	12	20^b	138
Nanoplate	H_2S	TiO_2	100	NA	300	NA	NA	5.5^b	139
Nanoplate	NO_2	WO_3	100	1	100	NA	NA	131.75^b	140
Nanoplate arrays	Xylene	α-MoO_3	100	NA	370	1	15	19.2^b	141
Thin film	NO_2	WO_3	1	NA	300	7	8	4.1^b	142
Thin film	Acetone	ZnO	100	NA	280	6	18	30^a	143
Thin film	2-Propanol	V_2O_5	5	NA	NA	3	10	176^a	144
Nanowall	NO_2	ZnO	50	NA	450	23	11	6.4^b	145

[Table is adapted from [146] The morphologies of the semiconductor oxides and their gas-sensing properties. Tingting Lin, Xin Lv, Shuang Li, and Qingji Wang. Sensors, 17(12):2779, (2017).]

$aS = R_a/R_g$; $^b S = R_g/R_a$; *LOD:* limit of detection; NA = not available.

at the interface. The 0D structures' sensing properties can be improved by assembling them into 3D hierarchical nanostructures. In the case of the 3D hierarchical nanostructures, connectivity between the particles offers high defect density. In general, the 1D nanostructures facilitate the fast transfer of electrons, and due to the high aspect ratio, it enhances sensing activity sites. 2D nanosheets also act as the building block for hierarchical nanostructures due to their large SVR. Diao et al. studied different 3D nanostructures: sphere-like and sisal-like structures for H_2S detection (Figure 4.10)] [147]. The response was found to be relatively high in the case

FIGURE 4.10 (a) FESEM image of sphere-like ZnO microstructure; (b) FESEM image of cauliflower-like ZnO microstructure; (c) FESEM image of sisal-like ZnO microstructure; (d) The response of different hierarchical ZnO microstructures sensors to 1000 ppb H_2S at different operating temperature (100–300°C) [Reprinted from [147] High response to H2S gas with facile synthesized hierarchical ZnO microstructures. Kaidi Diao, Minjie Zhou, Jicheng Zhang, Yongjian Tang, Shuxia Wang, and Xudong Cui. Sensors and Actuators B: Chemical, 219:30–37, (2015), with permission from Elsevier.]

of the spherical structures, which might be due to the particle-to-particle contacts. Microstructure-like spheres are covered with more nanowires, hence promoting particle-to-particle contact that motivates the change in the sensor's resistance. Recently it was found that 3D macro mesoporous exhibit superior gas sensing performance due to their highly porous structure and less agglomeration. This facilitates permeation of the gas and shortens gas diffusion length, which significantly enhances gas sensing performance. The channel between ZnO mesopores had an excellent sensing ability than the ZnO nanoparticles due to prolonged and torturous pathways in nanoparticles. The macropores in 3D ZnO nanostructures act as cavities and channels for the analyte molecules and provide the best response to even a very low concentration (100 ppm) of acetone.

4.5.2 Formation of the Stoichiometric Solid Solution

Host metals and dopants can form a solid solution. Formation of the solid solution can occur either by substituting the host atom from the lattice, or it could be present interstitially in the host lattice. In the case of substitution, the dopant or impurities can alter the host metal oxide's property if the size or charge differs from the host atom. However, interstitially present dopant atoms may generate strain in the lattice system; the electroneutrality condition is disturbed due to ionization. The formation of a solid solution can inhibit the particle growth, and thermal stability can be achieved [148]. The motion of the boundary between the particles controls the particle growth. The mean boundary velocity (MBV) can be expressed as follows:

$$v = M\,\Delta F \tag{4.22}$$

M is the mobility of the boundary and depends upon the diffusion. ΔF is the driving force. Thus, by reducing these two parameters, MBV can be reduced. Metastable solid solution forms a foreign cation segregation layer on the particle surface resulting in a reduction in both M and ΔF. This surface has an excess of cation, promotes inhibition of particle growth, and increases thermal stability. Moreover, this layer's formation decreases surface mobility and surface energy [149]. The linear relationship with dopant concentration and lattice parameter indicates solid solution formation. In a stoichiometric solid solution, if the charge of the dopant cation is less than the cation they will replace, then the dopant cation acts as the acceptor centre; however, if dopant charge is greater than replaced cations, then it acts as the donor centre [150]. In this solid solution, a cation interstitial or an oxygen vacancy can compensate for the extra negative charge as given by the following equations [150]:

$$A_2O = 2A_M + O_O + V_O \tag{4.23}$$

MO

$$A_2O = A_M + A_I + O_O \tag{4.24}$$

MO is the SMO with bivalent metal; D^{3+} and A^+ are dopant and acceptor dopant cations. However, to compensate for an extra positive charge in a stoichiometric solution, either oxygen interstitial or cation vacancy is needed, which can be expressed as follows [150]:

2MO

$$D_2O_3 = 2DM + OO + OI'' \tag{4.25}$$

(3MO)O

$$D_2O_3 = 2D_M + V_M'' + 3O_O \tag{4.26}$$

4.5.2.1 Change in Activation Energy

The relationship between activation energy and sensor response is expressed as follows:

$$R = AP_{02}^{-1/m}\, e^{-E/kT} \tag{4.27}$$

where R is the resistance, A is a constant, P_{02} is the partial pressure of oxygen, k is Bolzman constant, e is the activation energy, m is the defect-dependent constant, and T is the temperature. Taking logarithm of both sides:

$$\ln R = \ln A + \frac{E}{kT} - \frac{1}{m}\ln P_{o2} \tag{4.28}$$

If $P_{02} = 1$ at some two different temperatures, then the above equation becomes

$$E = k(T1 \times T2)/T_1T_2 \times (\ln R_1 \ln R_2) \tag{4.29}$$

where R_1 and R_2 are the response of the sensor at temperatures T_1 and T_2, respectively.

As the analyte gas molecules are absorbed in the sensing material's surface, they gain energy and get activated. Then these activated species react with adsorbed particles at the material's surface by the following reaction:

$$r = Cexp\left(-\frac{E}{kT}\right) \tag{4.30}$$

where r is termed as the rate of reaction, and C is called the pre-exponential factor. Rate of reaction constant increases with a decrease in activation energy. Thus, by lowering the activation energy, rate of response becomes higher. Some specific dopants such as platinum reported in the literature significantly affect the activation energy. This dopant also shows a spillover effect by providing active sites for analyte gases [151].

4.5.3 Generating Oxygen Vacancy

Oxygen vacancies act as the active sites for various adsorbates. Therefore, this type of point defect significantly affects many surface reactions. Oxygen vacancy can act as an electron donor site as well as a direct adsorption site. Sometimes oxygen vacancies can also act as the nucleation centre for the tiny metal clusters [152]. The variation in concentration of oxygen vacancies, creation, and annihilation of the oxygen as a function of the chemical potential of oxygen significantly affects the gas sensors' working principle. Doping of the metal oxide with dopant dominated by the oxygen vacancies or defect chemistry notably affects metal oxides' properties. The conductivity of the solid increases due to the presence of the oxygen vacancy. If the oxygen partial pressure is low, then more oxygen vacancies are expected to be generated. The determining factor for the generation of oxygen vacancies is ionic radii. The defect as the large anion or the cation cannot fit in the interstitial site available. In a literature report, it was found that initially up to 5 wt. % of aluminium-doped TiO_2, the crystalline size and particle size increase followed by a further decrease. Aluminium-doped titanium dioxide was very sensitive to carbon mono oxide and oxygen at a temperature of around 600°C compared to the pristine TiO_2-based sensors ~~(9)~~ [153].

4.5.4 Changes in the Electronic Structure/Bandgap

Based on the change in conductivity, the SMOs can be categorized into (i) bulk sensitive and (ii) surface sensitive. In the case of bulk-sensitive material, the creation of the Ti interstitial and oxygen vacancy results in conductivity change [154]. In ZnO and SnO_2, conductivity varies due to band bending. The surface oxygen species with negative charge introduces a band bending directed in an upward direction resulting in a reduction of conduction band charge carriers, which in turn cause a drop in the conductivity. The oxygen species' equilibrium with the gas species to be sensed decides the concentration of oxygen species. The reducing gases decrease the oxygen species' concentration by reacting with pre-adsorbed oxygen species, thus increasing the gas sensors' conductivity [154]. The surface conductivity is influenced by the donor and acceptor concentration, hence adsorbed oxygen vacancies, hydrogen atoms, and chemisorbed oxygen. The addition of external impurities in SMOs can overcome their limitations and improve their properties [155].

4.6 NOVEL SMO NANOSTRUCTURES AS GAS SENSOR

In the recent few decades, various nanostructures such as nanotubes, nanowires, nanoneedles, nanosheets, and many more nanostructures were explored for future electronic devices due to the high SVR of these structures. Among these structures, SMO nanowires are widely used to detect toxic gases. SMO nanowires have better gas-sensing abilities in comparison to the respective bulk and thin films [156, 157]. The metal oxide nanoparticles and other nanostructured base sensing devices are summarized in Table 4.3 and Table 4.4, respectively.

4.6.1 Electron Depletion

The SMOs used for gas-sensing applications are generally crystalline, and these crystallites are connected to each other by necks. Different grains are linked with large aggregates, which are connected by grain boundaries. The adsorbed O_2 molecules at surface grains trap conduction band electrons as ions at the surface, which results in the formation of EDL [188]. The relationship between sensitivity and grain size has already been discussed. The gaseous analyte's acceptor and donor electrons get adsorbed on the SMOs' surface and swap electrons with the sensing material. For an acceptor, the electron is collected from SMO leading to a reduction in

TABLE 4.3 Summary of Studies of Nanostructures SMO for Gas Sensing

Morphology	SMO	Gas	Reference
Nanowire	NiO	Ethanol, Acetone (500°C)	Kaur et al. (158)
Nanowire	WO_3	H_2 S	Krainer et al .(159)
Nanowire	ZnO, CuO	Ethanol	Zappa (160)
Nanotube	TiO_2	CO, Acetone, Ethanol (400°C, 500°C)	Galstyan et al.(161)
Core shell nanostructures	TiO_2..CuO	NO_2	Park et al.(162)
Nanosheets	CuO	Acetone, Methanol, Ethanol (200ppm, 340°C)	Umar et al.(163)

[Table is adapted from [18] Semiconductor metal oxide gas sensors: A review. Ananya Dey, Materials science, and Engineering: B, 229:206–217, (2018) with permission from Elsevier.]

electron concentration and sensitivity. On the contrary, in donor impurities, electrons are provided to the SMO leading to an enhancement in conductivity. The charge transfer between semiconductor metal oxide and adsorbed gaseous species affects gas sensors' conductivity. In polycrystalline SMO, this interaction occurs. The gas detection process using the metal oxides involves oxidation/reduction of semiconductors, the transport of the conduction band electrons between delocalized and localized states, direct adsorption of chemical species semiconductor surface, complex reactions between different adsorbed species on the surface, and catalytic effect [189].

4.6.2 Band Bending

At high temperatures, electrons smoothly flow through SMO film's grain boundaries. A potential barrier is formed at the SMO surface due to the absorbed oxygen. A charge layer forms due to SMO surface interaction with atmospheric oxygen, which traps electron bulk material. The repulsion force exerted by the charged oxygen species results in the formation of EDL, and barrier potential is increased. This constrains electron flow and increases resistance [190]. Below 147°C, oxygen is absorbed as O_2^- and between 147°C and 397°C, oxygen is ionosorbed as O^-, which is the main operating temperature for sensors. The donor site, which originates from intrinsic oxygen vacancies, provides electrons for this process. These conduction band electrons are extracted and get trapped, resulting in a space charge layer. The negative surface charges develop a potential barrier at the

TABLE 4.4 Summary of Studies of Nanoparticles SMO for Gas Sensing

Nano Particles	Properties of nanoparticles	Synthesis Techniques	Fabrication Technology	Sensing gas	Author
TiO_2	Size: 3 nm–30 nm	Chemically modified sol-gel technique	Dip-coating technique	Ethanol (100 ppm), Methanol (100 ppm), CO (100–300 ppm) NO2 (0.5–4 ppm) Temperature = 400 °C and 500 °C RH = 30%	Garzella et al. [164]
SnO_2	Size: 40 nm	NR*	Screen-printing	NH3 (937.5 ppm) Temperature = 104 °C–480 °C RH** = 20%	Llobet et al. [165]
CeO_2-TiO_2	NR	Sol–gel	Spin-coating technique	O2 (10,000 ppm) Temperature = 420 °C RH = < 5%	Trinchiet al. [166]
SnO_2	Size: 17.8 nm	Aerosol flame reactor (Premixed Flame)	Thick film deposition by drop coating method	NO2 (10–5000 ppb) Propanol (10–300 ppm) CO (500–10,000 ppm) Temperature = 200 °C–400 °C RH = 50%	Sham et al [167]
Nb doped TiO_2	Size:10–15 nm Specific area: 70–80 m²/g	Water – in-oil micro emulsion technique	Thick-film sensor	CO (1000 ppm) Temperature = 650 °C RH = NR	Anukunprasert et al [168]
F doped SnO_2	Size: 12–15 nm (calcinated at 550 °C) Surface area: 70 m²/g	Sol-gel	Micro-electromechanical system (MEMS)	CO, H2, C3H8, CH4 (100–600 ppm) Temperature = 22 °C, RH = 50%	Han et al. [169]
TiO_2	Size: 15 nm	NR*	Micro-electromechanical system, nanoparticles deposited spin coating	Methanol (50 ppm) Temperature = 375 °C–475 °C RH = dry air	Benkstein et al. [170]
Pt/SnO_2	Size: 10 nm	Flame spray pyrolysis (FSP)	In-situ deposition	CO (8–50 ppm) Temperature = 350 °C RH = dry air	Madler et al. [171]

TiO_2	Surface area:36–103 m2/g Size: 5–43 nm	Flame-spray Pyrolysis	Drop-coating method	Isoprene, ethanol, Acetone and CO (1–75 ppm) Temperature = 500 °C RH = dry ai	Telekiet al. [172]
Al doped TiO_2	Size:53 nm–76 nm	Citrate–Nitrate auto combustion method	NR*	CO (100–500 ppm) Temperature = 600 °C RH = dry air	Choi et al [153]
TiO_2	Size:53 nm–76 nm	NR*	DC reactive magnetron sputtering	NH3 (500 ppm) Temperature = 250 °C	Karunagaran et al. [173]
TiO_2	NR	Novel chemical route	Matrix assisted pulsed laser evaporation (MAPL E)	VOC (ethanol, acetone, 20–200 ppm) Temperature = 250 °C–500 °C, RH = dry air	Rella et al. [174]
CuO	Ave size: 10 nm	Thermal deposition method	NR*	NO2 (200 °C)	Li et al. [175]
Tungusten trioxide(WO_3)	NR	Acidification method	Nanoparticle paste is screenprinted on an alumina substrate equipped with a pair of comb-type Au microelectrode	NO2 (50–1000 ppb) Temperature = 200 °C RH = dry air	Kida et al. [176]
Tugusten trioxide nanotubes	Lamellar structured with diameter: 100–350 nm, thickness 20–50 nm, Specific surface area: 7–19 m2/g	Sol-gel	Nanotubes were fabricated on porous aluminium oxide membranes	NO_2, CH_3OH (0.2 μmol/mol, 2 μmol/ mol, 20 μmol/mol) Temperature = 200°C RH = dry air	Gerlitz et al. [177]
PdO loaded SnO_2 (nano - composites)	200 nm pore diameter & 600 nm pore length. Single nanotube is formed by discrete grain size of 60–80 nm	Reverse-Micelle method	Screen-Printing Method	CO Temperature = 300 °C RH = NR	Yuasa et al. [178]

(continued)

TABLE 4.4 (Continued)

Nano Particles	Properties of nanoparticles	Synthesis Techniques	Fabrication Technology	Sensing gas	Author
$TiO_2 - CeO_2$	Size: 7–12 nm	Sol–gel	NR*	CO (25–400 ppm) Temperature = 200 °C RH 30%	Mohammadi and Fray [179]
SnO_2	Ave grain size: 17 nm–28 nm	Flame aerosol reactor system	(Immediate power down) IPD mechanism	Ethanol vapor (100 ppm) Temperature = 250 °C RH = N	Zhan et al. [180]
Pt-TiO_2	NR	NR*	Screen-printing	H2, O2 Temperature = 500 °C–800 °C	Zhang et al [181]
TiO_2	NR*	Sol–gel	Dip-coating method	NH3 (56 ppm, 103 ppm, 156 ppm with changing background concentration of NO) Temperature = 350 °C RH = NR	Biskupski et al. [182]
TiO_2 nanotubes	Particle size: 7 nm–10 nm	NR*	Electrochemical anodization process	Formaldehyde (10–50 ppm) Room temperature RH = ~40%	Shiwai et al. [183]
CuO	NR*	Spray pyrolysis	NR*	H2S (250 °C)	Bari et al. [184]
NiO	29 nm	Chemical deposition and pyrolysis process	NR*	Benzaldehyde	Yang et al. [185]
In_2O_3	NR*	Film is grown by molecular beam epitaxy (MBE)	NR*	Ozone	Rombach et al [186]
In_2O_3	NR*	NR*	NR*	NO2 (50 ppb) 130 °C	Gonzalez et al [187]

[*Table is adapted from [18] Semiconductor metal oxide gas sensors:* A review. Ananya Dey, Materials Science, and Engineering: B, 229:206–217, (2018) with permission from Elsevier.]

**NR*: Not reported.

surface. The magnitude and width of this potential barrier are controlled by the surface charge and hence on the adsorbed oxygen [191].

4.6.3 Resistance Change

The adsorption of reducing gas molecules injects electrons into the conduction band of N-type semiconductors, and thus increases the conductivity; however, for p-type semiconductors, conductivity decreases. Thus, SMO gas sensors work as variable resistors that depend on the concentration of the analyte gas molecules [190]. In the case of oxidizing gases, the response phenomenon is reversed. Various binary oxides such as TiO_2, ZnO, and MoO_3, having bandgap ranging from 2 to 4 eV, generally show n-type character on the exposure of minority gas molecules under air ambient. In contrast, CuO, NiO, Co_3O_4, and Mn_3O_4 show p-type characteristics. Few of the oxides' sensing properties switch between p- and n-type characteristics under some set of conditions such as oxygen partial pressure, temperature, gas concentration, and so on. Both n- and p-type can have superior sensing abilities at different temperatures [192].

4.6.4 Incorporation of Oxide Heterojunction

With the evolution of material technology, a vast range of functional materials is now available to improve the sensors' efficiency. Similarly, several SMOs are currently being investigated to form p–n junctions to boost the performance of the devices. Under the exposure of target gases, the barrier height and width at the junction device's material interface get modified, resulting in more sensitivity [193]. As the oxide semiconductors are inherently doped (e.g., n-ZnO) due to their interstitial defects, their oppositely doped counterparts (e.g., p-ZnO) are relatively unstable. Hence, it is hard to grow a homojunction using oxide semiconductors. As a result, SMOs with different bandgaps are combined to realize heterojunction. The presence of non-identical bandgap, electron affinity, and band discontinuity at the interface allows the heterojunction to own unique advantages. There are three types of heterojunctions: n–n heterojunction, p–p heterojunction, and p–n heterojunction, as shown in Figure 4.11.

When an anisotype heterojunction is formed, electrons diffuse from n to p and holes move from p to n. Consequently, depletion regions form near the interface of p–n materials. When there is a flow of charged particles between the p and n regions to establish equilibrium, Fermi levels of the semiconductors become continuous through the semiconductor, and band bending occurs in conduction and valence bands at the depletion regions.

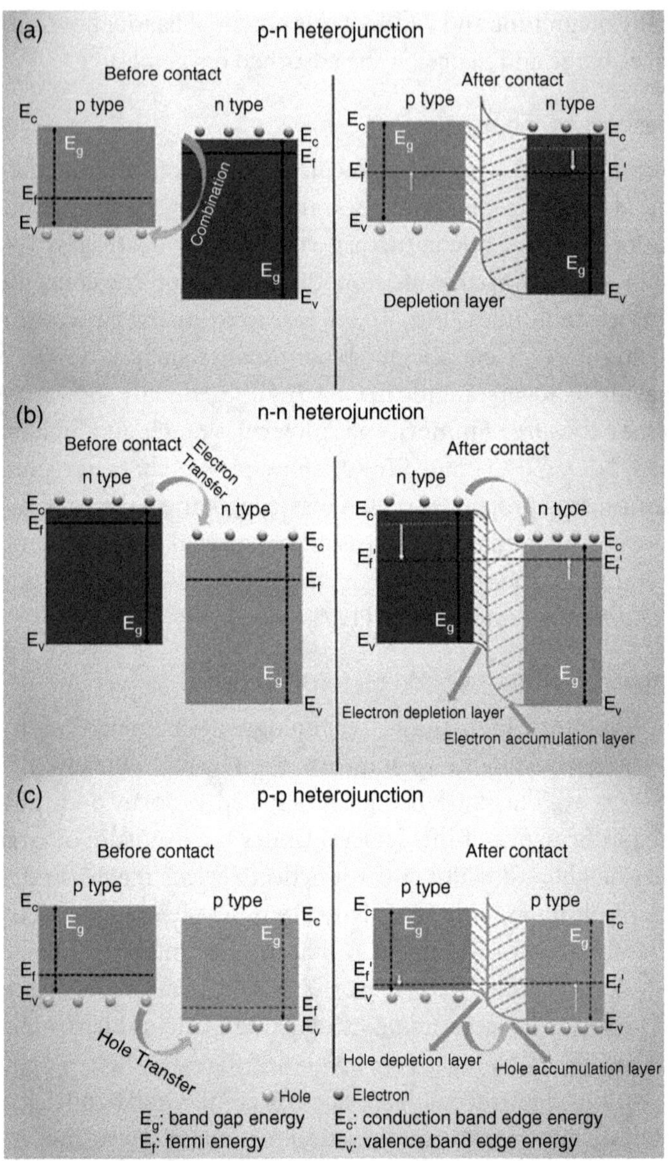

FIGURE 4.11 Schematic illustrations of the energy band structures at heterojunction interfaces of different types of heterojunctions: (a) p–n, (b) n–n, and (c) p–p. [Reprinted from [194] Advances in designs and mechanisms of semiconducting metal oxide, nanostructures for high-precision gas sensors operated at room temperature, Z. Li, H. Li, Z. Wu, M. Wang, J. Luo, H. Torun, P. Hu, C. Yang, M. Grundmann, X. Liu and Y. Fu, Mater. Horiz., 6, 470, (2019), with permission from Royal Society of Chemistry.]

Similarly, for isotype heterojunction, a band bending develops at the interface due to the diffusion of majority carriers from one side to another side (electrons and holes). A nanowire has a very high SVR leading to more gas absorption for a given volume. However, the processing routes to grow a single nanowire are expensive, and the resistance may vary from nanowire to nanowire leading to an issue in reproducibility. As a result, nanowire arrays and their associated heterojunction derivatives have been extensively investigated to enhance gas sensors' performance. For longitudinal p–n junctions, the adsorption of oxygen results in core-shell structured EDL and HAL formation. The subsequent reaction of these layers with target reducing/oxidizing gas is responsible for modulation in core-shell thickness and the corresponding variation in electrical properties of the junction. CuO–ZnO [195] and NiO–ZnO [196] are examples of this type of junction. However, when CuO–SnO junctions are exposed to H_2S gas, the high electron affinity of CuO results in the conversion of CuO surface into CuS and subsequent lowering of the junction barrier and resistance of the device [197]. Kim et al. revealed that p-type semiconductors can adsorb more O_2 than n-type semiconductors. Thus, under the exposure of C_2H_5OH gas, p-NiO nanoparticles loaded on the inside wall of n-SnO_2 hollow spheres exhibited an ultrafast recovery mechanism (<5 s) [198]. The minimum height of the VLS-grown single crystalline 1D nanowires is an order of magnitude more than the EDL thickness, which hinders the preparation of highly sensitive gas sensors [20]. This fact led to the development of core-shell nanostructures [199], nanoparticle embellished nanowires [200], and porous MoO_3/SnO_2 nanoflakes observed significant improvement in performance. Moreover, NiO nanoclusters' incorporation in SnO_2 nanowire boosted the sensor's gas selectivity due to its catalytic activities. Table 4.5 compares the attributes of some gas sensor devices.

4.7 PLASMONIC NANOPARTICLE-LOADED OXIDE NANOSTRUCTURES FOR GAS SENSING APPLICATIONS

Noble metal (Ag, Au, Pt) nanostructures exhibit distinct properties compared with their bulk counterparts under photon illumination. When the light wave is incident on metal nanoparticles, the electrons inside the NPs oscillate collectively. If the incident photon frequency matches with the frequency of oscillation, some exciting anomalies are observed in these nanostructures' optical properties termed surface plasmon resonance

TABLE 4.5 Comparison between Gas Sensor Devices

Material	Type	Growth	Target Gas	Concentration (ppm)	Temperature	Gas Response	Recovery Time	Ref
NiO/ZnO	p-n	Hydrothermal	Ethanol	100	200	16	6 s	201
PdO-WO_3	p-n	Low pressure chemical vapor deposition process, photolithography	Hydrogen	100	155	100	3 s	202
NiO/In_2O_3	p-n	Hydrothermal	Methanol	600	260	85.55	31.78 s	203
ZnO/CuO	p-n	Hydrothermal	Ethanol	50	29	9	420	204
NiO-ZnO	p-n	Hydrothermal	2-propanol	95	300	19.1	70 s and 55 s	205
NiO-SnO2	p-n	Electron beam evaporation and radio frequency magnetron sputtering	Ethanol	0.1	250	7.9	15 and 100	206
TiO_2–SnO_2	n-n	Polyol method	Ethanol	50	350	40	7 and 5	207
ZnO-WO_3	n-n	Hydrothermal	Ethanol	50	300	16.2		208
NiO/ Co_3O_4	p-p	Chemical synthesis followed by annealing	H2S	100	150	136		209
CuO-NiO	p-p	Hydrothermal	NO_2	100	Room Temperature		2 s	210

(SPR). SPR leads to very high photon absorption and scattering cross-section, localization of electric fields in the minimal area, and local heating effect in nanostructures. The plasmonic effect is applied in the fabrication of photovoltaic cells, UV–Vis photodetectors, cancer/tumour treatment, photocatalysis, and so on. Some research papers recently reported the incorporation of plasmonic nanomaterials in the active material to realize efficient gas sensor devices. Buso et al. conducted preparation of gold (Au)–NiO nanoparticle loaded SiO_2 substrates using sol-gel synthesis and dip-coating methods. In normal air, the nanostructures exhibited a strong absorbance band at 553 nm due to the plasmonic oscillation of electrons in Au NPs. However, under the exposure of target gas, the relative permittivity of the nanostructure ambience changes, modulating the appearance and frequency of the absorbance peak. As a result, the plasmon band in the absorption spectrum shifted by 22 nm (from 563 nm to 541 nm) in H_2 gas [211]. Similarly, Gaspera and his co-workers reported the Au NP functionalized TiO_2–NiO SMO composite by sol-gel method, spin coating, and annealing. In the open air, the sample exhibited a peak at 593 nm in its absorption spectrum attributed to Au SPR. But when the model is exposed to an H_2S environment, the charge transfer between Au and S atoms results in EDL at the Au surface, and the exciting oscillating s electrons decay through this. As a result, a broad and damped peak appears in the sample's absorption spectra [212].

4.8 CORE-SHELL NANOWIRE-BASED GAS SENSOR

The continuous progress in nanostructure growth mechanism has enabled the development of various novel nanostructures with efficient sensing properties. The core-shell nanowires receive more research interest than pristine nanowires due to their short carrier travel distance, higher junction area, ability to tune the scattering and absorption wavelength, and so on. Kim et al. reported the SnO_2–Cu_2O core-shell nanowire-based gas sensor and investigated the core-shell thickness's influence on response [213]. Although most of the electrons available in the thin Cu_2O shell layer depletion region modulated the resistance effectively under gas exposure, the smaller volume of shell compared to the core volume led to a diminished response. Moreover, the response varies inversely with the shell thickness. When shell thickness became close to the Debye length of MOS, the best performance was observed, as can be seen in Figure 4.12 [214].

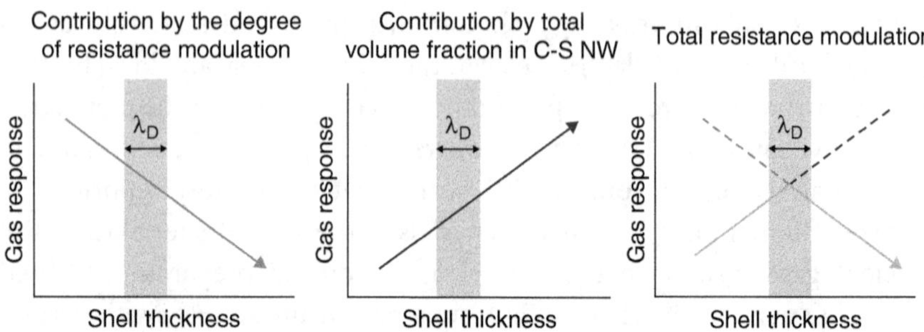

FIGURE 4.12 Total resistance modulation of the C-S NWs in the presence of reducing gases. [Reprinted from [194] Chemiresistive Sensing Behavior of SnO_2 (n)–Cu_2O (p) Core–Shell Nanowires, Jae-Hun Kim, Akash Katoch, Soo-Hyun Kim, and Sang Sub Kim, ACS Appl. Mater. Interfaces 2015, 7, 28, 15351–15358, with permission from American Chemical Society.]

4.9 LOW-POWER DESIGN TECHNIQUES

The gas sensors lack efficient response at room temperature. Application of high temperature or UV illumination can improve the sensitivity. The conventional method of placing a micro-heater in the sensor leads to increased power consumption. To overcome this problem, self-heating sensors are implemented, which reduce power consumption significantly. However, the inhomogeneous power dissipation may give rise to a large response time. Illumination of UV photons excites the electrons from valence to the conduction band. This increase in the active site in the semiconductor surface results in more adsorption and reaction of target molecules [19, 215–217].

4.10 OXIDE NANOSTRUCTURES FOR ELECTRONIC NOSE IMPLEMENTATION

Humans can distinguish different types of fragrances, odour, taste, and so on by smelling their characteristics. However, the human olfactory ability is limited by the total number of olfactory receptor cells (50 million) compared to the other animals (e.g., 200 million for a dog). As a result, the human nose cannot detect the presence of many harmful gases. Also, the inhalation of toxic gas can result in the deterioration of health conditions too. Hence, the gas sensor research also focuses on the development of an electronic nose system. An electronic nose's central functional units are a data sampling unit, gas sensor array, signal conditioning and preprocessing data block, and pattern recognition elements [218]. Under the exposure of gas, the chemical/physical properties of the SMOs are modulated, leading to a variation in the SMOs' conductivity. The corresponding variations in the electrical signals are processed, and the pattern recognition system predicts the gas identity. As compared to the sophisticated instruments, the electronic noses are cost-effective and offer bulk sample processing, fast response, and non-destructive diagnosis. Electronic noises are used in explosive detection, food quality investigation, environmental monitoring, and so forth.

4.11 CHALLENGES IN GAS SENSOR INTEGRATION IN TRADITIONAL CMOS TECHNOLOGY

Although the gas sensors find a wide range of applications in versatile domains, the systems' high price is a significant obstacle to these devices' rapid commercialization. For example, a solid-state gas sensor unit costs nearly ~10 $, but the cost of interfacing with an instrument using the drive circuit and signal processing unit raises to 500 $. The primary reasons for this high price are (1) very high manual manufacturing process cost, (2) the presence of discrete components in the interface PCB circuit, and (3) fireproof packaging. As a result, their applications are limited to commercial applications such as processing plants, power stations, and so on. Moreover, the power consumption level is increased by the microheater incorporated to improve sensor performance. Hence the integration in low-cost conventional microelectronic fabrication needs to be performed, eliminating the problems of weak selectivity, ageing, strong temperature, and humidity dependence [15].

4.12 CONCLUSION

This chapter presents an overview of the ongoing research in SMO nanostructure-based gas sensors. Their working principle is elaborated with gas adsorption/desorption, subsequent electron depletion/hole accumulation layer formation, and corresponding resistance variation models. The various FoMs for the sensor's performance and the factors affecting them were discussed. Also, the available growth techniques to fabricate nanostructures suitable for gas sensing purposes were illustrated. The structural and electronic modification of active materials and their role in enhancing the performance parameters were explained. Finally, the challenges in the integration of the gas sensor in the conventional CMOS platform are presented. The review may help the readers to explore new and novel materials with optimum morphology and electronic property to realize highly efficient gas sensors.

REFERENCES

1. Hao Cheng Ji, Wen Zeng, and Yanqiong Li. Gas sensing mechanisms of metal oxide semiconductors: a focus review. *Nanoscale*, 11(47):22664–22684, 2019.
2. Min-Ru Wu, Wei-Zhong Li, Chun-Yi Tung, Chiung-Yi Huang, Yi-Hung Chiang, Po-Liang Liu, and Ray-Hua Horng. No gas sensor based on znga2o4 epilayer grown by metalorganic chemical vapor deposition. *Scientific Reports*, 9(1):7459, 2019.
3. Rahul Kumar, Neeraj Goel, and Mahesh Kumar. High performance NO_2 sensor using mos2 nanowires network. *Applied Physics Letters*, 112(5), 2018.
4. Lauren McCauley (8 September 2016) "Making case for Clean air, World Bank says Pollution cost global economy $5 trillion." *Common Dreams*. Retrieved 3 February, 2018.
5. AA Mane and AV Moholkar. Orthorhombic moo3 nanobelts based NO_2 gas sensor. *Applied Surface Science*, 405:427–440, 2017.
6. Partha Bhattacharyya, PK Basu, Biplob Mondal, and H Saha. A low power mems gas sensor based on nanocrystalline zno thin films for sensing methane. *Microelectronics Reliability*, 48(11–12):1772–1779, 2008.
7. Salimeh Kimiagar, Vahid Najafi, Bartlomiej Witkowski, Rafal Pietruszka, and Marek Godlewski. High performance and low temperature coal mine gas sensor activated by UV-irradiation. *Scientific Reports*, 8(1):16298, 2018.
8. Jayasmita Jana, Mainak Ganguly, and Tarasankar Pal. Enlightening surface plasmon resonance effect of metal nanoparticles for practical spectroscopic application. *RSC Advances*, 6(89):86174–86211, 2016.
9. Eric G Barbagiovanni, David J Lockwood, Peter J Simpson, and Lyudmila V Goncharova. Quantum confinement in Si and Ge nanostructures: Theory and experiment. *Applied Physics Reviews*, 1(1), 2014.
10. Nirav Joshi, Takeshi Hayasaka, Yumeng Liu, Huiliang Liu, Osvaldo N Oliveira, and Liwei Lin. A review on chemiresistive room temperature gas sensors based on metal oxide nanostructures, graphene and 2d transition metal dichalcogenides. *Microchimica Acta*, 185:1–16, 2018.
11. Sapana Ranwa, Mohit Kumar, Jitendra Singh, Mattia Fanetti, and Mahesh Kumar. Schottky-contacted vertically self-aligned ZnO nanorods for hydrogen gas nanosensor applications. *Journal of Applied Physics*, 118(3), 2015.

12. Mohit Kumar, Vijendra Singh Bhati, and Mahesh Kumar. Effect of Schottky barrier height on hydrogen gas sensitivity of metal/TiO2 nanoplates. *International Journal of Hdrogen Energy*, 42(34):22082–22089, 2017.
13. Manjeet Kumar, Vishwa Bhatt, AC Abhyankar, Joondong Kim, Akshay Kumar, Sagar H Patil, and Ju-Hyung Yun. New insights towards strikingly improved room temperature ethanol sensing properties of p-type Ce-doped SnO2 sensors. *Scientific Reports*, 8(1):8079, 2018.
14. Balaji Jayaraman, Navakanta Bhat, and Rudra Pratap. Thermal characterization of microheaters from the dynamic response. *Journal of Micromechanics and Microengineering*, 19(8):085006, 2009.
15. Julian W Gardner, Prasanta K Guha, Florin Udrea, and James A Covington. CMOS interfacing for integrated gas sensors: A review. *IEEE Sensors Journal*, 10(12):1833–1848, 2010.
16. Jian Zhang, Ziyu Qin, Dawen Zeng, and Changsheng Xie. Metal-oxide-semiconductor based gas sensors: screening, preparation, and integration. *Physical Chemistry Chemical Physics*, 19(9):6313–6329, 2017.
17. Xiao Liu, Sitian Cheng, Hong Liu, Sha Hu, Daqiang Zhang, and HuanshengNing. A survey on gas sensing technology. *Sensors*, 12(7):9635–9665, 2012.
18. Ananya Dey. Semiconductor metal oxide gas sensors: A review. *Materials Science and Engineering: B*, 229:206–217, 2018.
19. Ali Mirzaei, Jae-Hyoung Lee, Sanjit Manohar Majhi, Matthieu Weber, Mikhael Bechelany, Hyoun Woo Kim, and Sang Sub Kim. Resistive gas sensors based on metal-oxide nanowires. *Journal of Applied Physics*, 126(24), 2019.
20. Hyo-Joong Kim and Jong-Heun Lee. Highly sensitive and selective gas sensors using p-type oxide semiconductors: Overview. *Sensors and Actuators B: Chemical*, 192:607–627, 2014.
21. Turja Nandy, Ronald A Coutu Jr, and Cristinel Ababei. Carbon monoxide sensing technologies for next-generation cyber-physical systems. *Sensors*, 18(10):3443, 2018.
22. Brian Yuliarto, Gilang Gumilar, and Ni Luh Wulan Septiani. SnO2 nanostructure as pollutant gas sensors: synthesis, sensing performances, and mechanism. *Advances in Materials Science and Engineering*, 2015, (2015):694823. https://doi.org/10.1155/2015/694823
23. Michael Hiibner, Cristian Eugen Simion, A Tomescu-Stanoiu, Suman Pokhrel, Nicolae Barsan, and Udo Weimar. Influence of humidity on co sensing with p-type cuo thick film gas sensors. *Sensors and Actuators B: Chemical*, 153(2):347–353, 2011.
24. James Welty, Gregory L Rorrer, and David G Foster. *Fundamentals of Momentum, Heat, and Mass Transfer*. John Wiley & Sons, 2020.

25. Svante Arrhenius. Uber die dissociationswarme und den einfluss der temperatur auf den dissociationsgrad der elektrolyte. *Zeitschrift fur physikalische Chemie*, 4(1):96–116, 1889.
26. G Korotcenkov and BK Cho. Metal oxide composites in conductometric gas sensors: Achievements and challenges. *Sensors and, Actuators B: Chemical*, 244:182–210, 2017.
27. VE Bochenkov, GB Sergeev, et al. Sensitivity, selectivity, and stability of gas-sensitive metal-oxide nanostructures. *Metal Oxide Nanostructures and Their Applications*, 3:31–52, 2010.
28. Marc J Madou and S Roy Morrison. *Chemical Sensing with Solid State Devices*. Elsevier, 2012.
29. Hisahito Ogawa, Atsushi Abe, Masahiro Nishikawa, and Shigeru Hayakawa. Electrical properties of tin oxide ultrafine particle films. *Journal of the Electrochemical Society*, 128(9):2020, 1981.
30. Hisahito Ogawa, Masahiro Nishikawa, and Atsushi Abe. Hall measurement studies and an electrical conduction model of tin oxide ultrafine particle films. *Journal of Applied Physics*, 53(6):4448–4455, 1982.
31. Chaonan Xu, Jun Tamaki, Norio Miura, and Noboru Yamazoe. Grain size effects on gas sensitivity of porous sno2-based elements. *Sensors and Actuators B: Chemical*, 3(2):147–155, 1991.
32. Nicolae Barsan and Udo Weimar. Conduction model of metal oxide gas sensors. *Journal of Electroceramics*, 7:143–167, 2001.
33. A Rothschild, Y Komem, and N Ashkenasy. Quantitative evaluation of chemisorption processes on semiconductors. *Journal of Applied Physics*, 92(12):7090–7097, 2002.
34. JW Orton and MJ Powell. The hall effect in polycrystalline and powdered semiconductors. *Reports on Progress in Physics*, 43(11):1263, 1980.
35. Yasuhiro Shimizu, Takeo Hyodo, and Makoto Egashira. Mesoporous semiconducting oxides for gas sensor application. *Journal of the European Ceramic Society*, 24(6):1389–1398, 2004.
36. Takeo Hyodo, Norihiro Nishida, Yasuhiro Shimizu, and Makoto Egashira. Preparation and gas-sensing properties of thermally stable mesoporous sno2. *Sensors and Actuators B: Chemical*, 83(1–3):209–215, 2002.
37. U-Sung Choi, Go Sakai, Kengo Shimanoe, and Noboru Yamazoe. Sensing properties of au-loaded sno2-co3o4 composites to co and h2. *Sensors and Actuators B: Chemical*, 107(1):397–401, 2005.
38. Wurzinger and G Reinhardt. Co-sensing properties of doped sno2 sensors in h2-rich gases. *Sensors and Actuators B: Chemical*, 103(1–2):104–110, 2004.
39. A Wisitsoraat, A Tuantranont, C Thanachayanont, V Patthanasettakul, and P Singjai. Electron beam evaporated carbon nanotube dispersed SnO2 thin film gas sensor. *Journal of Electroceramics*, 17:45–49, 2006.

40. Justin T McCue and Jackie Y Ying. Sno2- in2o3 nanocomposites as semiconductor gas sensors for Co and No x detection. *Chemistry of Materials*, 19(5):1009–1015, 2007.
41. Haining Ma, Lingmin Yu, Xiong Yuan, Yuan Li, Chun Li, Mingli Yin, and Xinhui Fan. Room temperature photoelectric NO2 gas sensor based on direct growth of walnut-like In2O3 nanostructures. *Journal of Alloys and Compounds*, 782:1121–1126, 2019.
42. Yonghui Deng. Interfacial interaction model between gas molecules and semiconducting metal oxides. In *Semiconducting Metal Oxides for Gas Sensing*, pages 189–252. Springer, 2023.
43. S Roy Morrison. Selectivity in semiconductor gas sensors. *Sensors and Actuators*, 12(4):425–440, 1987.
44. Hailin Tian, Huiqing Fan, Mengmeng Li, and Longtao Ma. Zeolitic imidazolate framework coated ZnO nanorods as molecular sieving to improve selectivity of formaldehyde gas sensor. *ACS Sensors*, 1(3):243–250, 2016.
45. Banglin Chen, Shengchang Xiang, and Guodong Qian. Metal-organic frameworks with functional pores for recognition of small molecules. *Accounts of Chemical Research*, 43(8):1115–1124, 2010.
46. Yuanjing Cui, Bin Li, Huajun He, Wei Zhou, Banglin Chen, and Guodong Qian. Metal-organic frameworks as platforms for functional materials. *Accounts of Chemical Research*, 49(3):483–493, 2016.
47. Rahul Kumar, Xianghong Liu, Jun Zhang, and Mahesh Kumar. Room-temperature gas sensors under photoactivation: from metal oxides to 2d materials. *Nano-Micro Letters*, 12:1–37, 2020.
48. Ashok B Gadkari, Tukaram J Shinde, and Pramod Nivrutti Vasambekar. Ferrite gas sensors. *IEEE Sensors Journal*, 11(4):849–861, 2010.
49. Nan Wu, Bing Wang, Cheng Han, Qiong Tian, Chunzhi Wu, Xiaoshan Zhang, Lian Sun, and Yingde Wang. Pt-decorated hierarchical sic nanofibers constructed by intertwined sic nanorods for high-temperature ammonia gas sensing. *Journal of Materials Chemistry C*, 7(24):7299–7307, 2019.
50. Arpan Kumar Nayak, Ruma Ghosh, Sumita Santra, Prasanta Kumar Guha, and Debabrata Pradhan. Hierarchical nanostructured WO3-SnO2 for selective sensing of volatile organic compounds. *Nanoscale*, 7(29):12460–12473, 2015.
51. Jing Li, Yijiang Lu, Qi Ye, Martin Cinke, Jie Han, and M Meyyappan. Carbon nanotube sensors for gas and organic vapor detection. *Nano Letters*, 3(7):929–933, 2003.
52. Yuliang Wang and Younan Xia. Bottom-up and top-down approaches to the synthesis of monodispersed spherical colloids of low melting-point metals. *Nano Letters*, 4(10):2047–2050, 2004.
53. a RS Wagner and s WC Ellis. Vapor-liquid-solid mechanism of single crystal growth. *Applied Physics Letters*, 4(5):89–90, 1964.

54. Won Il Park, D HI Kim, S-W Jung, and Gyu-Chul Yi. Metalorganic vapor-phase epitaxial growth of vertically well-aligned ZnO nanorods. *Applied Physics Letters*, 80(22):4232–4234, 2002.
55. Kurt W Kolasinski. Catalytic growth of nanowires: vapor-liquid-solid, vapor-solid-solid, solution-liquid-solid, and solid-liquid-solid growth. *Current Opinion in Solid State and Materials Science*, 10(3–4):182–191, 2006.
56. S Noor Mohammad. Self-catalysis: A contamination-free, substrate-free growth mechanism for single-crystal nanowire and nanotube growth by chemical vapor deposition. *The Journal of chemical physics*, 125(9), 2006.
57. S Noor Mohammad. Analysis of the vapor-liquid-solid mechanism for nanowire growth and a model for this mechanism. *Nano Letters*, 8(5):1532–1538, 2008.
58. Junqing Q Hu, Y Bando, and D Golberg. Self-catalyst growth and optical properties of novel sno2 fishbone-like nanoribbons. *Chemical Physics Letters*, 372(5–6):758–762, 2003.
59. Han, Wang, Jie, Wallace CH Choy, Yi Luo, TI Yuk, and JG Hou. Controllable synthesis and optical properties of novel ZnO cone arrays via vapor transport at low temperature. *The Journal of Physical Chemistry B*, 109(7):2733–2738, 2005.
60. XY Xue, YJ Chen, YG Liu, SL Shi, YG Wang, and TH Wang. Synthesis and ethanol sensing properties of indium-doped tin oxide nanowires. *Applied, Physics Letters*, 88(20), 2006.
61. AJ Chiquito, AJC Lanfredi, and ER Leite. One-dimensional character of Sn doped In2O3 nanowires probed by magnetotransport measurements. *Journal of Physics D: Applied Physics*, 41(4):045106, 2008.
62. S Hoffmann, J Bauer, C Ronning, Th Stelzner, J Michler, C Ballif, V Sivakov, and SH Christiansen. Axial pn junctions realized in silicon nanowires by ion implantation. *Nano Letters*, 9(4):1341–1344, 2009.
63. R Rakesh Kumar, K Narasimha Rao, K Rajanna, and AR Phani. Low temperature growth of SNO2 nanowires by electron beam evaporation and their application in UV light detection. *Materials Research Bulletin*, 48(4):1545–1552, 2013.
64. R Rakesh Kumar, D Yuvaraj, and K Narasimha Rao. Growth and characterization of germanium nanowires by electron beam evaporation. *Materials Letters*, 64(16):1766–1768, 2010.
65. R Rakesh Kumar, K Narasimha Rao, and AR Phani. Growth of silicon nanowires by electron beam evaporation using indium catalyst. *Materials Letters*, 66(1):110–112, 2012.
66. R Rakesh Kumar, K Narasimha Rao, and AR Phani. Bismuth catalyzed growth of silicon nanowires by electron beam evaporation. *Materials Letters*, 82:163–166, 2012.

67. Yutaka Ohya, Tomonori Yamamoto, and Takayuki Ban. Equilibrium dependence of the conductivity of pure and tin-doped indium oxide on oxygen partial pressure and formation of an intrinsic defect cluster. *Journal of the American, Ceramic Society*, 91(1):240–245, 2008.
68. Guojian Ren, Zhimeng Li, Weiting Yang, Muhammad Faheem, Jianbo Xing, Xiaoqin Zou, Qinhe Pan, Guangshan Zhu, and Yu Du. Zno@ zif-8 core-shell microspheres for improved ethanol gas sensing. *Sensors and Actuators B: Chemical*, 284:421–427, 2019.
69. Duk-Il Suh, Clare Chisu Byeon, and Chang-Lyoul Lee. Synthesis and optical characterization of vertically grown ZnO nanowires in high crystallinity through vapor-liquid-solid growth mechanism. *Applied Surface Science*, 257(5):1454–1456, 2010.
70. Saujan V Sivaram, Ho Yee Hui, Maria De La Mata, Jordi Arbiol, and Michael A Filler. Surface hydrogen enables subeutectic vapor-liquid-solid semiconductor nanowire growth. *Nano Letters*, 16(11):6717–6723, 2016.
71. Martin Ek and Michael A Filler. Atomic-scale choreography of vapor-liquid-solid nanowire growth. *Accounts of Chemical Research*, 51(1):118–126, 2018.
72. Chuanwei Cheng and Hong Jin Fan. Branched nanowires: synthesis and energy applications. *Nano Today*, 7(4):327–343, 2012.
73. Qing Wan, Jin Huang, Zhong Xie, Taihong Wang, Eric N Dattoli, and Wei Lu. Branched sno2 nanowires on metallic nanowire backbones for ethanol sensors application. *Applied Physics Letters*, 92(10), 2008.
74. Soyeon An, Sunghoon Park, Hyunsung Ko, Changhyun Jin, Wan In Lee, and Chongmu Lee. Enhanced Gas Sensing Properties of Branched ZnO Nanowires. *Thin Solid Films*, 547:241–245, 2013.
75. Ahmed Mohamed El-Toni, Mohamed A Habila, Joselito Puzon Labis, Zeid A ALOthman, Mansour Alhoshan, Ahmed A Elzatahry, and Fan Zhang. Design, synthesis and applications of core-shell, hollow core, and nanorattle multifunctional nanostructures. *Nanoscale*, 8(5):2510–2531, 2016.
76. Priyanka Karnati, Sheikh Akbar, and Patricia A Morris. Conduction mechanisms in one dimensional core-shell nanostructures for gas sensing: A review. *Sensors and Actuators B: Chemical*, 295:127–143, 2019.
77. Rajib Ghosh Chaudhuri and Santanu Paria. Core/shell nanoparticles: classes, properties, synthesis mechanisms, characterization, and applications. *Chemical Reviews*, 112(4):2373–2433, 2012.
78. Dario Zappa, Vardan Galstyan, Navpreet Kaur, Hashitha MM Munasinghe Arachchige, Orhan Sisman, and Elisabetta Comini. "metal oxide-based heterostructures for gas sensors"-a review. *Analytica Chimica Acta*, 1039:1–23, 2018.

79. Jani Hamalainen, Mikko Ritala, and Markku Leskela. Atomic layer deposition of noble metals and their oxides. *Chemistry of Materials*, 26(1):786–801, 2014.
80. Matthieu Weber, Emerson Coy, Igor Iatsunskyi, Luis Yate, Philippe Miele, and Mikhael Bechelany. Mechanical properties of boron nitride thin films prepared by atomic layer deposition. *Cryst Eng Comm*, 19(41):6089–6094, 2017.
81. Matthieu Weber, Igor Iatsunskyi, Emerson Coy, Philippe Miele, David Cornu, and Mikhael Bechelany. Novel and facile route for the synthesis of tunable boron nitride nanotubes combining atomic layer deposition and annealing processes for water purification. *Advanced Materials Interfaces*, 5(16):1800056, 2018.
82. Christophe Detavernier, Jolien Dendooven, Sreeprasanth Pulinthanathu Sree, Karl F Ludwig, and Johan A Martens. Tailoring nanoporous materials by atomic layer deposition. *Chemical Society Reviews*, 40(11):5242–5253, 2011.
83. Matthieu Weber, Jin-Young Kim, Jae-Hyoung Lee, Jae-Hun Kim, Igor Iatsunskyi, Emerson Coy, Philippe Miele, Mikhael Bechelany, and Sang Sub Kim. Highly efficient hydrogen sensors based on Pd nanoparticles supported on boron nitride coated ZnO nanowires. *Journal of Materials Chemistry A*, 7(14):8107–8116, 2019.
84. Li-Yuan Zhu, Kaiping Yuan, Jian-Guo Yang, Hong-Ping Ma, Tao Wang, Xin-Ming Ji, Ji-Jun Feng, Anjana Devi, and Hong-Liang Lu. Fabrication of heterostructured p-cuo/n-sno2 core-shell nanowires for enhanced sensitive and selective formaldehyde detection. *Sensors and Actuators B: Chemical*, 290:233–241, 2019.
85. M Tonezzer and NV Hieu. Size-dependent response of single-nanowire gas sensors. *Sensors and Actuators B: Chemical*, 163(1):146–152, 2012.
86. Young-Jin Choi, In-Sung Hwang, Jae-Gwan Park, Kyoung Jin Choi, Jae-Hwan Park, and Jong-Heun Lee. Novel fabrication of an SnO2 nanowire gas sensor with high sensitivity. *Nanotechnology*, 19(9):095508, 2008.
87. Oleg Lupan, VV Ursaki, Guangyu Chai, Lee Chow, Gennady A Emelchenko, IM Tiginyanu, Alex N Gruzintsev, and AN Redkin. Selective hydrogen gas nanosensor using individual ZnO nanowire with fast response at room temperature. *Sensors and Actuators B: Chemical*, 144(l):56–66, 2010.
88. Daihua Zhang, Zuqin Liu, Chao Li, Tao Tang, Xiaolei Liu, Song Han, Bo Lei, and Chongwu Zhou. Detection of no2 down to ppb levels using individual and multiple in2o3 nanowire devices. *Nano Letters*, 4(10):1919–1924, 2004.
89. Mohammad Zhian Asadzadeh, Anton Kock, Maxim Popov, Stephan Steinhauer, Jtirgen Spitaler, and Lorenz Romaner. Response modeling of single

SnO2 nanowire gas sensors. *Sensors and Actuators B: Chemical*, 295:22–29, 2019.

90. LC Tien, HT Wang, BS Kang, F Ren, PW Sadik, DP Norton, SJ Pearton, and Jenshan Lin. Room-temperature hydrogen-selective sensing using single Pt-coated ZnO nanowires at microwatt power levels. *Electrochemical and Solid-State Letters*, 8(9):G230, 2005.
91. Jae Hoon Bang, Myung Sik Choi, Ali Mirzaei, Yong Jung Kwon, Sang Sub Kim, Tae Whan Kim, and Hyoun Woo Kim. Selective NO2 sensor based on bi2o3 branched sno2 nanowires. *Sensors and Actuators B: Chemical*, 274:356–369, 2018.
92. Sang Sub Kim, Han Gil Na, Sun-Woo Choi, Dong Sub Kwak, and Hyoun Woo Kim. Novel growth of CuO-functionalized, branched SnO2 nanowires and their application to H2S sensors. *Journal of Physics D: Applied Physics*, 45(20):205301, 2012.
93. Hyoun Woo Kim, Han Gil Na, Yong Jung Kwon, Hong Yeon Cho, and Chongmu Lee. Decoration of Co nanoparticles on ZnO-branched SnO2 nanowires to enhance gas sensing. *Sensors and Actuators B: Chemical*, 219:22–29, 2015.
94. Hyung-Sik Woo, Chang-Hoon Kwak, Jae-Ho Chung, and Jong-Heun Lee. Co-doped branched ZnO nanowires for ultraselective and sensitive detection of xylene. *ACS Applied, Materials & Interfaces*, 6(24):22553–22560, 2014.
95. Hyung-Sik Woo, Chang-Hoon Kwak, Jae-Ho Chung, and Jong-Heun Lee. Highly selective and sensitive xylene sensors using Ni-doped branched ZnO nanowire networks. *Sensors and Actuators B: Chemical*, 216:358–366, 2015.
96. Jiaqiang Xu, Qingyi Pan, Zhizhuang Tian, et al. Grain size control and gas sensing properties of ZnO gas sensor. *Sensors and Actuators B: Chemical*, 66(1–3):277–279, 2000.
97. Wenhu Tan, Xiaofan Ruan, Qiuxiang Yu, Zetai Yu, and Xintang Huang. Fabrication of a SnO2-based acetone gas sensor enhanced by molecular imprinting. *Sensors*, 15(l):352–364, 2014.
98. Rui Zhang, Wei Pang, Zhihong Feng, Xuejiao Chen, Yan Chen, Qing Zhang, Hao Zhang, Chongling Sun, J Joshua Yang, and Daihua Zhang. Enabling selectivity and fast recovery of ZnO nanowire gas sensors through resistive switching. *Sensors and Actuators B: Chemical*, 238:357–363, 2017.
99. Mingli Yin, Lingmin Yu, and Shengzhong Liu. Synthesis of thickness-controlled cuboid WO3 nanosheets and their exposed facets-dependent acetone sensing properties. *Journal of Alloys and Compounds*, 696:490–497, 2017.

100. Fengrui Li, Jikang Jian, Rong Wu, Jin Li, and Yanfei Sun. Synthesis, electrochemical and gas sensing properties of In2O3 nanostructures with different morphologies. *Journal of Alloys and Compounds*, 645:178–183, 2015.
101. Kuan Tian, Xiao-Xue Wang, Hua-Yao Li, Reddeppa Nadimicherla, and Xin Guo. Lotus pollen derived 3-dimensional hierarchically porous NiO microspheres for NO2 gas sensing. *Sensors and Actuators B: Chemical*, 227:554–560, 2016.
102. Xiaoying Peng, Zhongming Wang, Pan Huang, Xun Chen, Xianzhi Fu, and Wenxin Dai. Comparative study of two different TiO2 film sensors on response to H2 under UV light and room temperature. *Sensors*, 16(8):1249, 2016.
103. Herbert Gleiter. Nanostructured materials: basic concepts and microstructure. *Acta Materialia*, 48(1):1–29, 2000.
104. VV Pokropivny and VV Skorokhod. Classification of nanostructures by dimensionality and concept of surface forms engineering in nanomaterial science. *Materials Science and Engineering: C*, 27(5–8):990–993, 2007.
105. Ankur Gupta, Tamilselvan Sakthivel, and Sudipta Seal. Recent development in 2d materials beyond graphene. *Progress in Materials Science*, 73:44–126, 2015.
106. Takafumi Akamatsu, Toshio Itoh, Noriya Izu, and Woosuck Shin. No and no2 sensing properties of wo3 and co3o4 based gas sensors. *Sensors*, 13(9):12467–12481, 2013.
107. Sanjeev K Bhardwaj, Neha Bhardwaj, Manil Kukkar, Amit L Sharma, Ki-Hyun Kim, and Akash Deep. Formation of high-purity indium oxide nanoparticles and their application to sensitive detection of ammonia. *Sensors*, 15(12):31930–31938, 2015.
108. Wenlong Zhang, Bin Yang, Jingquan Liu, Xiang Chen, Xiaolin Wang, and Chunsheng Yang. Highly sensitive and low operating temperature SnO2 gas sensor doped by Cu and Zn two elements. *Sensors and Actuators B: Chemical*, 243:982–989, 2017.
109. Qingji Wang, Fangmeng Liu, Jun Lin, and Geyu Lu. Gas-sensing properties of in-sn oxides composites synthesized by hydrothermal method. *Sensors and Actuators B: Chemical*, 234:130–136, 2016.
110. Haitao Gao, He Jia, Benedikt Bierer, Jtirgen Wollenstein, Yan Lu, and Stefan Palzer. Scalable gas sensors fabrication to integrate metal oxide nanoparticles with well-defined shape and size. *Sensors and Actuators B: Chemical*, 249:639–646, 2017.
111. Lili Wang, Rui Zhang, Tingting Zhou, Zheng Lou, Jianan Deng, and Tong Zhang. Concave cu2o octahedral nanoparticles as an advanced sensing material for benzene (c6h6) and nitrogen dioxide (no2) detection. *Sensors and Actuators B: Chemical*, 223:311–317, 2016.

112. Haoxiong Yu, Zhilong Song, Qian Liu, Xiao Ji, Jianqiao Liu, Songman Xu, Hao Kan, Baohui Zhang, Jingyao Liu, Jianjun Jiang, et al. Colloidal synthesis of tungsten oxide quantum dots for sensitive and selective H2S gas detection. *Sensors and Actuators B: Chemical*, 248:1029–1036, 2017.
113. Bingxin Xiao, Fei Wang, Chengbo Zhai, Pan Wang, Chuanhai Xiao, and Mingzhe Zhang. Facile synthesis of In2O3 nanoparticles for sensing properties at low detection temperature. *Sensors and Actuators B: Chemical*, 235:251–257, 2016.
114. Rupesh S Devan, Ranjit A Patil, Jin-Han Lin, and Yuan-Ron Ma. One-dimensional metal-oxide nanostructures: recent developments in synthesis, characterization, and applications. *Advanced Functional Mat*erials, 22(16):3326–3370, 2012.
115. Ahsanulhaq Qurashi, Toshinari Yamazaki, EM El-Maghraby, and Toshio Kikuta. Fabrication and gas sensing properties of in2o3 nanopushpins. *Applied Physics Letters*, 95(15), 2009.
116. Xianghua Kong and Yadong Li. High sensitivity of Cuo modified SnO2 nanoribbons to H2S at room temperature. *Sensors and Actuators B: Chemical*, 105(2):449–453, 2005.
117. MM Arafat, B Dinan, Sheikh A Akbar, and ASMA Haseeb. Gas sensors based on one dimensional nanostructured metal-oxides: a review. Sensors, 12(6):7207–7258, 2012.
118. Z Ying, Q Wan, ZT Song, and SL Feng. SnO2 nanowhiskers and their ethanol sensing characteristics. *Nanotechnology*, 15(11):1682, 2004.
119. Peiguang Hu, Guojun Du, Weijia Zhou, Jingjie Cui, Jianjian Lin, Hong Liu, Duo Liu, Jiyang Wang, and Shaowei Chen. Enhancement of ethanol vapor sensing of tio2 nanobelts by surface engineering. *ACS Applied Materials & Interfaces*, 2(11):3263–3269, 2010.
120. Zhenhuan Zhao, Jian Tian, Yuanhua Sang, Andreu Cabot, and Hong Liu. Structure, synthesis, and applications of TiO2 nanobelts. *Advanced, Materials*, 27(16):2557–2582, 2015.
121. Fang Fang, Lu Bai, Dongsheng Song, Hongping Yang, Xiaoming Sun, Hongyu Sun, and Jing Zhu. Ag-modified In2O3/ZnO nanobundles with high formaldehyde gas-sensing performance. *Sensors*, 15(8):20086–20096, 2015.
122. Siyuan Feng-Chen, Ali Aldalbahi, and Peter Xianping Feng. Nanostructured tungsten oxide composite for high-performance gas sensors. *Sensors*, 15(10):27035–27046, 2015.
123. Hyung-Sik Woo, Chan Woong Na, and Jong-Heun Lee. Design of highly selective gas sensors via physicochemical modification of oxide nanowires: overview. *Sensors*, 16(9):1531, 2016.
124. J Yu, KW Cheung, WH Yan, YX Li, and D Ho. High-sensitivity low-power tungsten doped niobium oxide nanorods sensor for nitrogen

dioxide air pollution monitoring. *Sensors and Actuators B: Chemical*, 238:204–213, 2017.

125. Changhui Zhao, Jinglong Bai, Baoyu Huang, Yaling Wang, Jinyuan Zhou, and Erqing Xie. Grain refining effect of calcium dopants on gas-sensing properties of electrospun a-Fe2O3 nanotubes. *Sensors and Actuators B: Chemical*, 231:552–560, 2016.
126. Navpreet Kaur, Mandeep Singh, and Elisabetta Comini. One-dimensional nanostructured oxide chemoresistive sensors. *Langmuir*, 36(23):6326–6344, 2020.
127. Shuang Yang, Yueli Liu, Wen Chen, Wei Jin, Jing Zhou, Han Zhang, and Galina S Zakharova. High sensitivity and good selectivity of ultra-long MoO3 nanobelts for trimethylamine gas. *Sensors and Actuators B: Chemical*, 226:478–485, 2016.
128. Ji-Wook Yoon, Hyo-Joong Kim, Hyun-Mook Jeong, and Jong-Heun Lee. Gas sensing characteristics of p-type Cr2O3 and Co3O4 nanofibers depending on inter-particle connectivity. *Sensors and Actuators B: Chemical*, 202:263–271, 2014.
129. Yuhai Dou, Jiantie Xu, Boyang Ruan, Qiannan Liu, Yuede Pan, Ziqi Sun, and Shi Xue Dou. Atomic layer-by-layer Co3O4/graphene composite for high performance lithium-ion batteries. *Advanced, Energy Materials*, 6(8):1501835, 2016.
130. Hui Miao, Guowei Zhang, Xiaoyun Hu, Jianglong Mu, Tongxin Han, Jun Fan, Changjun Zhu, Lixun Song, Jintao Bai, and Xun Hou. A novel strategy to prepare 2d g-C3N4 nanosheets and their photoelectrochemical properties. *Journal of Alloys and Compounds*, 690:669–676, 2017.
131. S Narasimman, L Balakrishnan, SR Meher, R Sivacoumar, and ZC Alex. Influence of surface functionalization on the gas sensing characteristics of ZnO nanorhombuses. *Journal of Alloys and Compounds*, 706:186–197, 2017.
132. A Nijamudheen and Alexey V Akimov. Excited-state dynamics in two-dimensional heterostructures: Sir/TiO2 and ger/TiO2 (r= h, me) as promising photocatalysts. *Journal of Physical Chemistry C*, 121(12):6520–6532, 2017.
133. Jean Rouquerol, David Avnir, Craig W Fairbridge, Douglas H Everett, JM Haynes, Nicola Pernicone, John DF Ramsay, Kenneth Stafford William Sing, and Klaus K Unger. Recommendations for the characterization of porous solids (technical report). *Pure and Applied, Chemistry*, 66(8):1739–1758, 1994.
134. Xue Wang, Juan Su, Hui Chen, Guo-Dong Li, Zhifang Shi, Haifeng Zou, and Xiaoxin Zou. Ultrathin In2O3 nanosheets with uniform mesopores for highly sensitive nitric oxide detection. *ACS Applied, Materials & Interfaces*, 9(19):16335–16342, 2017.

135. Zhijie Li, Zhijie Lin, Ningning Wang, Junqiang Wang, Wei Liu, Kai Sun, Yong Qing Fu, and Zhiguo Wang. High precision NH3 sensing using network nano-sheet Co3O4 arrays based sensor at room temperature. *Sensors and Actuators B: Chemical*, 235:222–231, 2016.
136. German Escalante, Hector Juarez, and Paloma Fernandez. Characterization and sensing properties of ZnO film prepared by single source chemical vapor deposition. *Advanced Powder Technology*, 28(1):23–29, 2017.
137. Ziyue Zhang, Zhen Wen, Zhizhen Ye, and Liping Zhu. Gas sensors based on ultrathin porous Co3O4 nanosheets to detect acetone at low temperature. *RSC Advances*, 5(74):59976–59982, 2015.
138. Chuanhai Xiao, Tianye Yang, Mingyan Chuai, Bingxin Xiao, and Mingzhe Zhang. Synthesis of zno nanosheet arrays with exposed (100) facets for gas sensing applications. *Physical Chemistry Chemical Physics*, 18(1):325–330, 2016.
139. Wei Guo, Qingqin Feng, Yufang Tao, Lijun Zheng, Zongyao Han, and Jianmin Ma. Systematic investigation on the gas-sensing performance of TiO2 nanoplate sensors for enhanced detection on toxic gases. *Materials Research Bulletin*, 73:302–307, 2016.
140. SS Shendage, VL Patil, SA Vanalakar, SP Patil, NS Harale, JL Bhosale, JH Kim, and PS Patil. Sensitive and selective NO2 gas sensor based on Wo3 nanoplates. *Sensors and Actuators B: Chemical*, 240:426–433, 2017.
141. Haiyu Qin, Yali Cao, Jing Xie, Hui Xu, and Dianzeng Jia. Solid-state chemical synthesis and xylene-sensing properties of a-MoO3 arrays assembled by nanoplates. *Sensors and Actuators B: Chemical*, 242:769–776, 2017.
142. A Maity and SB Majumder. No2 sensing and selectivity characteristics of tungsten oxide thin films. *Sensors and Actuators B: Chemical*, 206:423–429, 2015.
143. F Teimoori, K Khojier, and NZ Dehnavi. Investigation of sensitivity and selectivity of ZnO thin film to volatile organic compounds. *Journal of Theoretical and Applied Physics*, 11:157–163, 2017.
144. Pranav Shashidhar Karthikeyan, P Dhivya, P Deepak Raj, and M Sridharan. V2o5 thin film for 2-propanol vapor sensing. *Materials Today: Proceedings*, 3(6):1510–1516, 2016.
145. Lingmin Yu, Fen Guo, Sheng Liu, Bing Yang, Yanxing Jiang, Lijun Qi, and Xinhui Fan. Both oxygen vacancies defects and porosity facilitated NO2 gas sensing response in 2d ZnO nanowalls at room temperature. *Journal of Alloys and Compounds*, 682:352–356, 2016.
146. Tingting Lin, Xin Lv, Shuang Li, and Qingji Wang. The morphologies of the semiconductor oxides and their gas-sensing properties. *Sensors*, 17(12):2779, 2017.
147. Kaidi Diao, Minjie Zhou, Jicheng Zhang, Yongjian Tang, Shuxia Wang, and Xudong Cui. High response to H2S gas with facile synthesized

hierarchical ZnO microstructures. *Sensors and Actuators B: Chemical*, 219:30–37, 2015.

148. AP Maciel, Paulo Noronha Lisboa-Filho, ER Leite, CO Paiva-Santos, WH Schreiner, Y Maniette, and Elson Longo. Microstructural and morphological analysis of pure and ce-doped tin dioxide nanoparticles. *Journal of the European Ceramic Society*, 23(5):707–713, 2003.
149. Edson R Leite, Adeilton P Maciel, Ingrid T Weber, Paulo N Lisboa-Filho, Elson Longo, CO Paiva-Santos, AVC Andrade, CA Pakoscimas, Y Maniette, and Wido H Schreiner. Development of metal oxide nanoparticles with high stability against particle growth using a metastable solid solution. *Advanced Materials*, 14(12):905–908, 2002.
150. DM Smyth. The effects of dopants on the properties of metal oxides. *Solid State Ionics*, 129(1–4):5–12, 2000.
151. Neeraj Goel, Kishor Kunal, Aditya Kushwaha, and Mahesh Kumar. Metal oxide semiconductors for gas sensing. *Engineering Reports*, 5(6):e12604, 2023.
152. Stefan Wendt, Renald Schaub, Jesper Matthiesen, Ebbe K Vestergaard, E Wahlstrom, Maria Dall Rasmussen, Peter Thostrup, LM Molina, Erik Laegsgaard, Ivan Stensgaard, et al. Oxygen vacancies on TiO2 (1 1 0) and their interaction with H2O and O2: A combined high-resolution STM and DFT study. *Surface Science*, 598(1–3):226–245, 2005.
153. Young Jin Choi, Zachary Seeley, Amit Bandyopadhyay, Susmita Bose, and Sheikh A Akbar. Aluminum-doped TiO2 nano-powders for gas sensors. *Sensors and Actuators B: Chemical*, 124(1):111–117, 2007.
154. Matthias Batzill and Ulrike Diebold. Surface studies of gas sensing metal oxides. *Physical Chemistry Chemical Physics*, 9(19):2307–2318, 2007.
155. Soumyadip Basu, and Palash Kumar Basu. Nanocrystalline metal oxides for methane sensors: role of noble metals. *Journal of Sensors*, 2009 (2009): 861968:1–861968:20.
156. E Brunet, Thomas Maier, Giorgio Cataldo Mutinati, Stephan Steinhauer, A Kock, Christian Gspan, and Werner Grogger. Comparison of the gas sensing performance of SnO2 thin film and SnO2 nanowire sensors. *Sensors and Actuators B: Chemical*, 165(1):110–118, 2012.
157. Chu Manh Hung, Dang Thi Thanh Le, and Nguyen Van Hieu. On-chip growth of semiconductor metal oxide nanowires for gas sensors: A review. *Journal of Science: Advanced Materials and Devices*, 2(3):263–285, 2017.
158. Navpreet Kaur, Elisabetta Comini, Nicola Poli, Dario Zappa, and Giorgio Sberveglieri. Nickel oxide nanowires growth by VLS technique for gas sensing application. *Procedia Engineering*, 120:760–763, 2015.
159. Johanna Krainer, Marco Deluca, Eva Lackner, Robert Wimmer-Teubenbacher, Florentyna Sosada, Christian Gspan, Karl Rohracher,

Ewald Wachmann, and Anton Kock. Cmos integrated tungsten oxide nanowire networks for ppb-level H2S sensing. *Procedia Engineering*, 168:272–275, 2016.

160. Dario Zappa, Elisabetta Comini, and Giorgio Sberveglieri. Gas-sensing properties of thermally-oxidized metal oxide nanowires. *Procedia Engineering*, 47:430–433, 2012.
161. V Galstyan, E Comini, C Baratto, M Ferroni, N Poli, G Faglia, E Bontempi, M Brisotto, and G Sberveglieri. Two-phase titania nanotubes for gas sensing. *Procedia Engineering*, 87:176–179, 2014.
162. Sunghoon Park, Soohyun Kim, Gun-Joo Sun, Wan In Lee, Kyoung Kook Kim, and Chongmu Lee. Fabrication and NO2 gas sensing performance of TeO 2-core/CuO-shell heterostructure nanorod sensors. *Nanoscale Research Letters*, 9:1–7, 2014.
163. Ahmad Umar, AA Alshahrani, H Algarni, and Rajesh Kumar. CuO nanosheets as potential scaffolds for gas sensing applications. *Sensors and Actuators B: Chemical*, 250:24–31, 2017.
164. Carlotta Garzella, Elisabetta Comini, E Tempesti, C Frigeri, and Giorgio Sberveglieri. TiO2 thin films by a novel sol-gel processing for gas sensor applications. *Sensors and Actuators B: Chemical*, 68(1–3):189–196, 2000.
165. E Llobet, P Ivanov, X Vilanova, J Brezmes, J Hubalek, K Malysz, I Gracia, C Cane, and X Correig. Screen-printed nanoparticle tin oxide films for high-yield sensor microsystems. *Sensors and Actuators B: Chemical*, 96(1–2):94–104, 2003.
166. Adrian Trinchi, YX Li, W Wlodarski, S Kaciulis, L Pandolfi, S Viticoli, Elisabetta Comini, and Giorgio Sberveglieri. Investigation of sol-gel prepared CeO2-TIO2 thin films for oxygen gas sensing. *Sensors and Actuators B: Chemical*, 95(1–3):145–150, 2003.
167. T Sahm, L Madler, A Gurlo, N Barsan, Sotiris E Pratsinis, and U Weimar. Flame spray synthesis of tin dioxide nanoparticles for gas sensing. *Sensors and actuators B: Chemical*, 98(2–3):148–153, 2004.
168. T Anukunprasert, C Saiwan, and Enrico Traversa. The development of gas sensor for carbon monoxide monitoring using nanostructure of Nb-TIO2. *Science and Technology of Advanced Materials*, 6(3–4):359–363, 2005.
169. Chi-Hwan Han, Sang-Do Han, Ishwar Singh, and Thierry Toupance. Micro-bead of nano-crystalline f-doped SnO2 as a sensitive hydrogen gas sensor. *Sensors and Actuators B: Chemical*, 109(2):264–269, 2005.
170. Kurt D Benkstein and Stephen Semancik. Mesoporous nanoparticle TiO2 thin films for conductometric gas sensing on microhotplate platforms. *Sensors and Actuators B: Chemical*, 113(1):445–453, 2006.

171. Lutz Madler, Albert Roessler, Sotiris E Pratsinis, Thorsten Sahm, Aleksander Gurlo, Nicolae Barsan, and Udo Weimar. Direct formation of highly porous gas-sensing films by in situ thermophoretic deposition of flame-made pt/sno2 nanoparticles. *Sensors and Actuators B: Chemical*, 114(1):283–295, 2006.
172. Alexandra Teleki, Sotiris E Pratsinis, K Kalyanasundaram, and PI Gouma. Sensing of organic vapors by flame-made TIO2 nanoparticles. *Sensors and Actuators B: Chemical*, 119(2):683–690, 2006.
173. B Karunagaran, Periyayya Uthirakumar, SJ Chung, S Velumani, and E-K Suh. Tio2 thin film gas sensor for monitoring ammonia. *Materials Characterization*, 58(8–9):680–684, 2007.
174. R Rella, J Spadavecchia, MG Manera, S Capone, A Taurino, Maurizio Martino, Anna Paola Caricato, and T Tunno. Acetone and ethanol solid-state gas sensors based on TiO2 nanoparticles thin film deposited by matrix assisted pulsed laser evaporation. *Sensors and Actuators B: Chemical*, 127(2):426–431, 2007.
175. Yueming Li, Jing Liang, Zhanliang Tao, and Jun Chen. CuO particles and plates: synthesis and gas-sensor application. *Materials Research Bulletin*, 43(8–9):2380–2385, 2008.
176. Tetsuya Kida, Aya Nishiyama, Masayoshi Yuasa, Kengo Shimanoe, and Noboru Yamazoe. Highly sensitive NO2 sensors using lamellar-structured Wo3 particles prepared by an acidification method. *Sensors and Actuators B: Chemical*, 135(2):568–574, 2009.
177. Reit Artzi-Gerlitz, Kurt D Benkstein, David L Lahr, Joshua L Hertz, Christopher B Montgomery, John E Bonevich, Steve Semancik, and Michael J Tarlov. Fabrication and gas sensing performance of parallel assemblies of metal oxide nanotubes supported by porous aluminum oxide membranes. *Sensors and Actuators B: Chemical*, 136(1):257–264, 2009.
178. Masayoshi Yuasa, Takanori Masaki, Tetsuya Kida, Kengo Shimanoe, and Noboru Yamazoe. Nano-sized PDO loaded SnO2 nanoparticles by reverse micelle method for highly sensitive co gas sensor. *Sensors and Actuators B: Chemical*, 136(1):99–104, 2009.
179. MR Mohammadi and DJ Fray. Nanostructured TiO2-CeO2 mixed oxides by an aqueous sol-gel process: Effect of Ce:Ti molar ratio on physical and sensing properties. *Sensors and Actuators B: Chemical*, 150(2):631–640, 2010.
180. Zili Zhan, Wei-Ning Wang, Liying Zhu, Woo-Jin An, and Pratim Biswas. Flame aerosol reactor synthesis of nanostructured SnO2 thin films: High gas-sensing properties by control of morphology. *Sensors and Actuators B: Chemical*, 150(2):609–615, 2010.

181. Maolin Zhang, Zhanheng Yuan, Jianping Song, and Cheng Zheng. Improvement and mechanism for the fast response of a pt/tio2 gas sensor. *Sensors and Actuators B: Chemical*, 148(1):87–92, 2010.
182. Diana Biskupski, Bettina Herbig, Gerhard Schottner, and Ralf Moos. Nanosized titania derived from a novel sol-gel process for ammonia gas sensor applications. *Sensors and Actuators B: Chemical*, 153(2):329–334, 2011.
183. Shiwei Lin, Dongrong Li, Jian Wu, Xiaogan Li, and SA Akbar. A selective room temperature formaldehyde gas sensor using TiO2 nanotube arrays. *Sensors and Actuators B: Chemical*, 156(2):505–509, 2011.
184. Ramesh H Bari, Sharad B Patil, and Anil R Bari. Spray-pyrolized nanostructured CuO thin films for H2S gas sensor. *International Nano Letters*, 3(1):12, 2013.
185. Fuchao Yang and Zhiguang Guo. Engineering NiO sensitive materials and its ultra-selective detection of benzaldehyde. *Journal of Colloid and Interface Science*, 467:192–202, 2016.
186. Julius Rombach, Oliver Bierwagen, Alexandra Papadogianni, Mischo Mischo, Volker Cimalla, Theresa Berthold, Stefan Krischok, and Marcel Himmerlich. Electrical conductivity and gas-sensing properties of Mg-doped and undoped single-crystalline In2O3 thin films: bulk vs. surface. *Procedia Engineering*, 120:79–82, 2015.
187. Oriol Gonzalez, Sergio Roso, Raul Calavia, Xavier Vilanova, and Eduard Llobet. NO2 sensing properties of thermally or UV activated In2O3 nano- octahedra. *Procedia Engineering*, 120:773–776, 2015.
188. Yu-Feng Sun, Shao-Bo Liu, Fan-Li Meng, Jin-Yun Liu, Zhen Jin, Ling-Tao Kong, and Jin-Huai Liu. Metal oxide nanostructures and their gas sensing properties: a review. *Sensors*, 12(3):2610–2631, 2012.
189. Ghenadii Korotcenkov. Metal oxides for solid-state gas sensors: What determines our choice? *Materials Science and Engineering: B*, 139(1):1–23, 2007.
190. Sofian M Kanan, Oussama M El-Kadri, Imad A Abu-Yousef, and Marsha C Kanan. Semiconducting metal oxide based sensors for selective gas pollutant detection. *Sensors*, 9(10):8158–8196, 2009.
191. Marion E Franke, Tobias J Koplin, and Ulrich Simon. Metal and metal oxide nanoparticles in chemiresistors: does the nanoscale matter? *Small*, 2(1):36–50, 2006.
192. Patrick T Moseley. Progress in the development of semiconducting metal oxide gas sensors: A review. *Measurement Science and Technology*, 28(8):082001, 2017.
193. Yingying Jian, Wenwen Hu, Zhenhuan Zhao, Pengfei Cheng, Hossam Haick, Mingshui Yao, and Weiwei Wu. Gas sensors based on

chemi-resistive hybrid functional nanomaterials. *Nano-Micro Letters*, 12:1–43, 2020.

194. Zhijie Li, Hao Li, Zhonglin Wu, Mingkui Wang, Jingting Luo, Hamdi Torun, PingAn Hu, Chang Yang, Marius Grundmann, Xiaoteng Liu, et al. Advances in designs and mechanisms of semiconducting metal oxide nanostructures for high-precision gas sensors operated at room temperature. *Materials Horizons*, 6(3):470–506, 2019.
195. Seymen Aygfin and David Cann. Hydrogen sensitivity of doped CuO/ZnO heterocontact sensors. *Sensors and Actuators B: Chemical*, 106(2):837–842, 2005.
196. Chan Woong Na, Hyung-Sik Woo, and Jong-Heun Lee. Design of highly sensitive volatile organic compound sensors by controlling NiO loading on ZnO nanowire networks. *RSC Advances*, 2(2):414–417, 2012.
197. Jun Tamaki, Kengo Shimanoe, Yoshihiro Yamada, Yoshifumi Yamamoto, Norio Miura, and Noboru Yamazoe. Dilute hydrogen sulfide sensing properties of CuO-SnO2 thin film prepared by low-pressure evaporation method. *Sensors and Actuators B: Chemical*, 49(1–2):121–125, 1998.
198. Hae-Ryong Kim, Kwon-Il Choi, Kang-Min Kim, Il-Doo Kim, Guozhong Cao, and Jong-Heun Lee. Ultra-fast responding and recovering C2H5 OH sensors using SnO2 hollow spheres prepared and activated by Ni templates. *Chemical Communications*, 46(28):5061–5063, 2010.
199. Xiaoxin Li, Xiaogan Li, Ning Chen, Xinye Li, Jianwei Zhang, Jun Yu, Jing Wang, and Zhenan Tang. Cuo-In2O3 core-shell nanowire based chemical gas sensors. *Journal of Nanomaterials*, 2014:8–8, 2014.
200. Chan Woong Na, Seung-Young Park, Jae-Ho Chung, and Jong-Heun Lee. Transformation of ZnO nanobelts into single-crystalline Mn3O4 nanowires. *ACS Applied Materials & Interfaces*, 4(12):6565–6572, 2012.
201. Hailin Tian, Huiqing Fan, Guangzhi Dong, Longtao Ma, and Jiangwei Ma. NiO/ZnO p-n heterostructures and their gas sensing properties for reduced operating temperature. *RSC Advances*, 6(110):109091–109098, 2016.
202. Seungmin Kwak, Young-Seok Shim, Yong Kyoung Yoo, Jin-Hyung Lee, Inho Kim, Jinseok Kim, Kyu Hyoung Lee, and Jeong Hoon Lee. Mems-based gas sensor using PdO-decorated TiO2 thin film for highly sensitive and selective H2 detection with low power consumption. *Electronic Materials Letters*, 14:305–313, 2018.
203. Hou Xuemei, Sun Yukun, and Bai Bo. Fabrication of cubic p-n heterojunction-like NiO/In2O3 composite microparticles and their enhanced gas sensing characteristics. *Journal of Nanomaterials*, 2016 (2016): 7589028. https://doi.org/10.1155/2016/7589028

204. PP Subha and MK Jayaraj. Enhanced room temperature gas sensing properties of low temperature solution processed ZnO/CuO heterojunction. *BMC Chemistry*, 13:1–11, 2019.
205. Sayan Dey, Swati Nag, Sumita Santra, Samit Kumar Ray, and Prasanta Kumar Guha. Voltage-controlled NiO/ZnO p-n heterojunction diode: A new approach towards selective VOC sensing. *Microsystems & Nanoengineering*, 6(1):35, 2020.
206. Jiabin Fang, Yiping Zhu, Dajun Wu, Chi Zhang, Shaohui Xu, Dayuan Xiong, Pingxiong Yang, Lianwei Wang, and Paul K Chu. Gas sensing properties of NiO/SnO2 heterojunction thin film. *Sensors and Actuators B: Chemical*, 252:1163–1168, 2017.
207. Tao Zhao, Pengpeng Qiu, Yuchi Fan, Jianping Yang, Wan Jiang, Lianjun Wang, Yonghui Deng, and Wei Luo. Hierarchical branched mesoporous TiO2-SnO2 nanocomposites with well-defined n-n heterojunctions for highly efficient ethanol sensing. *Advanced Science*, 6(24):1902008, 2019.
208. Yuan-Chang Liang and Che-Wei Chang. Improvement of ethanol gas-sensing responses of zno-wo3 composite nanorods through annealing induced local phase transformation. *Nanomaterials*, 9(5):669, 2019.
209. Kuan-Wei Chen, Ju-Heng Tsai, and Chun-Hua Chen. NiO functionalized Co3O4 hetero-nanocomposites with a novel apple-like architecture for Co gas sensing applications. *Materials Letters*, 255:126508, 2019.
210. He Xu, Jiawei Zhang, Afrasiab Ur Rehman, Lihong Gong, Kan Kan, Li Li, and Keying Shi. Synthesis of NiO@ CuO nanocomposite as high-performance gas sensing material for NO2 at room temperature. *Applied Surface Science*, 412:230–237, 2017.
211. D Buso, G Busato, M Guglielmi, A Martucci, V Bello, G Mattei, P Mazzoldi, and ML Post. Selective optical detection of H2 and Co with SiO2 sol-gel films containing NiO and Au nanoparticles. *Nanotechnology*, 18(47):475505, 2007.
212. Enrico Della Gaspera, Massimo Guglielmi, Stefano Agnoli, Gaetano Granozzi, Michael L Post, Valentina Bello, Giovanni Mattei, and Alessandro Martucci. Au nanoparticles in nanocrystalline Tio2-NiO films for SPR-based, selective H2S gas sensing. *Chemistry of Materials*, 22(11):3407–3417, 2010.
213. Jae-Hun Kim, Akash Katoch, Soo-Hyun Kim, and Sang Sub Kim. Chemiresistive sensing behavior of SnO2 (N)-Cu2O (p) core-shell nanowires. *ACS Applied Materials & Interfaces*, 7(28):15351–15358, 2015.
214. Jae-Hun Kim, Akash Katoch, and Sang Sub Kim. Optimum shell thickness and underlying sensing mechanism in p-n CuO-ZnO core-shell nanowires. *Sensors and Actuators B: Chemical*, 222:249–256, 2016.

215. P Camagni, G Faglia, P Galinetto, C Perego, G Samoggia, and G Sberveglieri. Photosensitivity activation of SnO2 thin film gas sensors at room temperature. *Sensors and Actuators B: Chemical*, 31(1–2):99–103, 1996.
216. Ehsan Espid and Fariborz Taghipour. UV-led photo-activated chemical gas sensors: A review. *Critical Reviews in Solid State and Materials Sciences*, 42(5):416–432, 2017.
217. Trinh Minh Ngoc, Nguyen Van Duy, Chu Manh Hung, Nguyen Due Hoa, Hugo Nguyen, Matteo Tonezzer, and Nguyen Van Hieu. Self-heated Ag-decorated SnO_2 nanowires with low power consumption used as a predictive virtual multisensor for H2S-selective sensing. *Analytica Chimica Acta*, 1069:108–116, 2019.
218. Khalil Arshak, E Moore, Gerard M Lyons, John Harris, and Seamus Clifford. A review of gas sensors employed in electronic nose applications. *Sensor Review*, 24(2):181–198, 2004.

CHAPTER 5

Study of Biosensors Based on Metal Oxide Semiconductor Nanostructures

Priyanka Chetri, Jay Chandra Dhar,
Ashok Palepu, and Michael Cholines Pedapudi

5.1 INTRODUCTION

Biosensor-based research has exponential enhancement over the last two decades. A biosensor is generally defined as a monitoring device that converts a biological response into a measurable and processable physical, chemical, or electrical signal with the help of a transducer. Developing a specific and cost-effective biosensor is more useful for accurate diagnosis and customized medicine. Biosensors can respond to the biologically active components like food, body fluids, cell cultures, enzymes, antibodies, hormone samples, etc., and the analyte can be glucose, urea, drug, or pesticide whose existence as well as the concentration should be measured. The challenging task while designing biosensors is to convert the biological samples information to a managed electronic signal because of the difficulty of relating electronic devices to the biological environment [1]. But in the case of electrochemical biosensors, the analysis is done with the content of a biological sample with direct conversion of a biological event to an electronic signal. The performance of the biosensor is determined

DOI: 10.1201/9781003464211-5

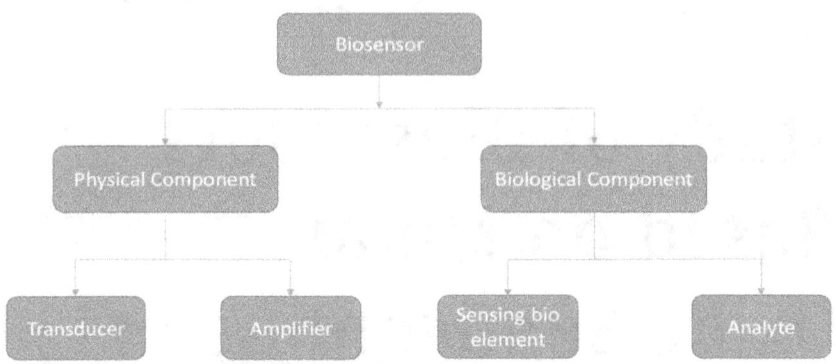

FIGURE 5.1 Schematic scheme of typical biosensor components.

by the stability of the immobilized enzyme on to the transducer. Thus, by harnessing the inherent specificity and sensitivity of biological molecules, biosensors offer rapid, accurate, and portable detection capabilities.

The typical scheme of a biosensor, reported in Figure 5.1, includes a biological component where the recognition of the sensing element is performed and thereafter it is able to bind the target analyte with the help of cell cultures, human samples (blood, urine, saliva, etc.), food samples, and environmental samples (air, water soil, etc.). An electrochemical active interface is being created for biological recognition process. A transducer which can convert a biochemical reaction into a processable signal depends upon the type of biosensor when it comes to interacting with biologically sensitive material like electrical charge, potential, current, temperature, measuring weight, and optical activity [2]. These can be used to measure the concentration of sugars, amino acids, alcohols, lipids, nucleotides, and others which can be specifically identified, and their concentration can be measured by these devices. Thereafter the amplifier amplifies the power of the measured signal for a useful output much higher than the input signal with less error detection. Biosensors have three important components: (1) recognition element, (2) transducer unit, and (3) signal processing unit.

To construct a successful biosensor several conditions must be considered; they are:

The biocatalyst must be very specific for the purpose of the analysis; it should be steady under normal storage conditions and show a low variation between examinees. The reaction should be as independent as manageable of such physical parameters as stirring, pH, and temperature [3]. The response should be accurate, precise, reproducible, and linear over the concentration range of interest, without dilution or concentration. If the biosensor is to be used for invasive monitoring in clinical situations, the probe must be tiny and biocompatible, having no toxic or antigenic effects [4]. For rapid measurements of analytes from human samples, it is desirable that the biosensor can provide real-time analysis. The complete biosensor should be cheap, small, portable, and capable of being used by semi-skilled operators.

5.2 BIOSENSOR TECHNOLOGIES: WORKING PRINCIPLE AND CHARACTERISTICS

5.2.1 Working Principle

The basic working principle of a biosensor is the specific and efficient interaction between a biological recognition element and the target analyte, which leads to a measurable signal generated by a transducer. This reaction changes the physicochemical properties of the transducer surface. This leads to a change in the optical/electrical properties of the transducer surface. The transducer converts the biochemical changes into an electrical, optical, thermal, or mass-related signal that can be easily detected and quantified. However, the bio element and the sensing element can be varied depending on the user requirement and its necessity as given in Figure 5.2.

Biosensors are generally highly selective due to the possibility of tailoring the specific interaction of compounds by immobilizing biological

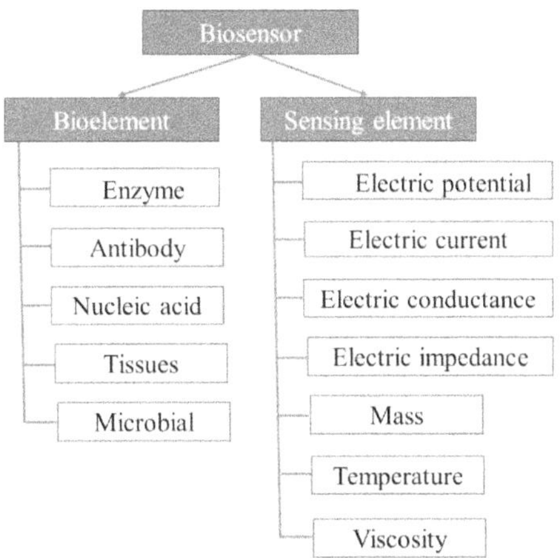

FIGURE 5.2 Classification of biosensor based on sensing element.

recognition elements on the sensor substrate that has a specific binding affinity to the desired molecule [5, 6]. Typical recognition elements used in biosensors are enzymes, nucleic acids, antibodies, tissues, and receptors. The whole area of biosensors started with the introduction of the first-generation glucose oxidase biosensor as exemplified by the glucose sensor. Electrochemical biosensors do not suffer the drawback of high sensor setup complexity and cost advantages. Electrochemical biosensors are robust, easy to miniaturize, have excellent detection limits, also with small analytic volumes, and the ability to be used in turbid biofluids with optically absorbing and fluorescing compounds. Electrochemical biosensors have suffered from a lack of surface architectures allowing high enough sensitivity and unique identification of the response with the desired biochemical event [7].

5.2.2 Characteristics of Biosensors

Biosensor is characterized by the following capabilities:

Linearity: It is the ability of a sensor which should be high for the detection of a high concentration of substrate.

Selectivity: This is the capability of biosensors to select and to measure only the desired biochemical species with minimal interference from many other species present in the environment.

Sensitivity: It is the value of electrode response per substrate concentration.

Detection Limit: It defines the minimum concentration of analyte to be measured.

Response Time: Kinetics of the reaction is the deciding factor for the response time of a biosensor. Enzymatic and immunological reactions are characterized by fast kinetics.

Lifetime: The lifetime of a biosensor is dependent on the method of immobilization of the biosensing element. With time, degenerative processes will be observed in the biosensing element and may cause the loss of the original sensing function.

To improve the quality of life in many areas such as environmental monitoring, health care, and food safety diagnosis, biosensors will play a major part. In the past few decades, nanomaterial-based biosensors have experienced rapid developments. The following discussion will focus on the nanomaterial-based biosensors.

5.2.3 Electrochemical Biosensors

An electrochemical biosensor is a self-contained integrated device that can provide specific quantitative or semi-quantitative analytic information using a biological recognition element that is engaged in direct spatial contact with an electrochemical transduction element. It is capable of directly converting a biological event to a variation of an electric quantity (current, potential, resistivity, capacitance, and impedance), which is measured by means of an electrode (transducer) in electrochemical biosensor [8]. The produced electric signal is proportional to the detection interaction between the analyte and target and to the concentration of the analyte.

The electrochemical cell is the best method to explain the working principle of an electrochemical biosensor. The electrochemical cell is composed of a working electrode, a reference electrode, and a counter electrode immersed in the electroactive species solution. The reaction of interest takes place on the working electrode. This electrode can be fabricated from different materials such as glassy carbon, platinum, gold, mercury, screen-printed materials, carbon paste, indium tin oxide-coated glass, silver, and semiconductors depending on the specific application [9]. In the reference electrode, a constant potential functional nanostructured interface for environmental and biomedical applications is generated.

The reference electrode should have low impedance, and it should be a nonpolarizable electrode to minimize the variation of the potential due to the passage of current. Half-cell components with high stability, with precise values of activity, and with fixed and reproducible potentials are needed for the fabrication of the reference electrode. The counter electrode is commonly fabricated with platinum, glass, graphite, carbon, and gold. The surface area of the counter electrode must be higher than the working electrode's area.

5.2.4 Potentiometric Transducers

A potentiometric biosensor measures a potential between an electrode and a reference electrode at zero current, generated by a biological detection reaction that leads to the alteration of a redox potential or to the activation of an ion. When ramp voltage is applied to an electrode in solution, a current flow occurs because of electrochemical reactions, the measured parameters are oxidation and reduction potential [10]. A potentiometric biosensor is composed of a perm-selective film and a surface sensitive to a bioactive material. The catalyzed reaction generated by the bioactive material (enzyme) consumes a particulate species of biomolecule that is detected by an ion-selective electrode. Extremely small concentration changes are detected by this technique, because of the logarithmic nature of the concentration response.

5.2.5 Amperometric Transducers

In amperometric transducers, the current produced by an electroactive analyte by means of its oxidation or reduction at an electrode surface is measured under a constant voltage applied to the working electrode. The applied potential compels the species to lose or gain electrons. The measured current is typical of the detection interaction and then is proportional to the rate of the electron transfer and to the concentration of the analyte [11]. The Clark electrode is the simplest type of amperometric biosensor. The bulk concentration of the electroactive species is proportional to the current peak value measured at a constant potential. Despite certain protein analytes, the sensitivity of amperometric devices is higher than potentiometric devices.

5.2.6 Optical Biosensors

Optical biosensors are based on the interaction of a biorecognition element with the optical field. They use two different techniques of recognition: label-free method, where the detected signal is generated directly by the interaction of the analyzed material with the transducer; and the label-based method, where the use of a label is necessary and the optical signal is generated by a colorimetric, fluorescent, or luminescent method. Simple molecules can be detected by enzymatic oxidation using label-assisted sensing, but labeling can alter the binding properties and therefore introduce systematic error to the biosensor analysis [12]. An optical biosensor produces a signal that is related to the concentration of an analyte and is composed of a miniaturized device containing a biodetector integrated with an optical-based transducer. Optical biosensors can be classified into different families, depending on the transduction method used, such as surface plasmon resonance (SPR), evanescent wave fluorescence, optical waveguide interferometry, bioluminescent optical fiber, optical waveguide interferometry, reflectometric interference spectroscopy, and surface-enhanced Raman scattering.

5.2.7 Piezoelectric Biosensors

Piezoelectric materials are widely used in many technological applications and are frequently embedded in electronics. Over the past 20 years, the use of these materials for biosensing applications has been investigated, proving them to be well-suitable for the fabrication of biosensors [13]. Electroacoustic devices offer a promising technology platform for the development of biosensors with high sensitivity and portability. In particular, the use of microfabrication techniques for acoustic wave-based biosensors led to device miniaturization, low power consumption, and complementary metal-oxide semiconductor compatibility. Electroacoustic devices, using the inverse piezoelectric effect, are based on the propagation of high-frequency acoustic waves in the bulk or on the surface of a medium [14]. Changes in the velocity and the amplitude are caused by mass loading or by the viscosity of the medium where the device is immersed. The conventional piezoelectric microbiosensor is based on the interaction of a bioanalyte with the sensitive layer deposited on the sensor surface. Mass or viscosity modification induced by analyte–sensing layer interaction can be detected by measuring changes in attenuation, velocity, delay time, and resonant frequency of the acoustic wave by electronic oscillators [15]. The fabrication processes are based on conventional microfabrication techniques such as the deposition of thin metal films, photolithography, wet and dry etching, and sputtering of thin films.

5.3 METAL OXIDE-BASED NANOSTRUCTURES

Nanomaterials possess unique structural, mechanical, chemical, and optoelectronic properties in comparison to bulk or macroscopic materials. These nanomaterials tend to offer properties such as quantum confinement effect, surface effect, small size effect, etc. which greatly promote a wide range of applications in medicine, electronics, biomaterials, environmental science, energy production, and biosensors. These nanomaterials have many advantages over the traditional material–modified electrodes. Firstly, the nanotextured surfaces offer an increased surface-area-to-volume ratio for the immobilization on the electrode which offers high sensitivity for detection of target analytes in their ambient environment. Secondly, some nanomaterials act as promoters, accelerating the electron transfer rate. Thirdly, some biocompatible nanomaterials can help proteins or cells to maintain their activities on the electrode for a long period providing biocompatibility.

Different synthesized nanomaterials such as nanoparticles, nanorods, nanowires, etc. are employed to effectively improve the characteristics of a biosensor performance [16–18]. However, the purpose and the use of an electrode remain the same, but the strategy to tune the properties of electrodes has sparked attention from researchers worldwide. Modification of electrode surfaces can be engineered as it provides an opportunity to control and tune the performance of nano biosensor platforms such as conferring selectivity, resisting fouling, eliminating nonspecific interactions, and thus improving the S/N ratio [19]. Tailoring the electrode surface via nanostructuring provides enhanced charge transport and steady-state diffusion in electrochemical biosensors. Doping is also an important aspect of increasing the sensitivity and selectivity of the bio-electrode. Carbon-based products (metal, ceramic, polymeric, metal oxide nanoparticles) were also studied in detail since these are abundant in nature, low cost, and chemically stable [20].

5.3.1 Thin Film Biosensors

Thin film-based biosensor is made from either organic and/or inorganic materials, such as metals, glass, polymers, silicon, or metal oxides. These are comprised of complex structures of thin films, which give enormous functionalities to the sensors. The working principle is based on the selective adsorption of analyte molecules on a functionalized thin film which act as physicochemical (optical, mechanical, magnetic, or electrical) transducers, converting the signal from the recognition of the biological analyte into another form of a measurable signal. The most crucial part of biosensors is the analyte-sensitive layer, that is, the layer which reacts to the biomolecule. On this surface, analyte-specific reaction starts, and therefore the atomic interactions, surface-free energies, and forces are different from the bulk of the material. Surface activation immobilizes the biological analytes on sensitive layers and for this purpose, self-assembled monolayers (SAM)-enabled surface modification techniques are the most widely used [21, 22]. Thin films can be used to functionalize (e.g., control hydrophobicity, bio affinity, biocompatibility, and electrical activity) the surface of biosensors with or without using surface treatment processes [23]. Some of the thin film dielectric coatings such as SiO_2, TiO_2, Al_2O_3, etc. are employed to obtain superior electrical, optical, magnetic, and mechanical properties. Even metallic thin films were also used to form either bond pads for electrical probing of the analyte or as activation surfaces for appropriate functionalization. To improve the adhesion between the active layers, metals such as Cr and Ti have been widely used underlying the Si substrates. Various works on organic thin films are also undergone where the process essentially consists of depositing a thin layer of organic films such as a variety of thiol molecules, proteins, and nucleic acid molecules [24, 25]. However, the selection of the surface for the uniform formation of the film depends on the activated surface material.

5.3.2 From TF to 1D

One-dimensional (1D) nanostructures such as nanowires, nanotubes, nanobelts, and nanosprings are the smallest dimension structures that can be used for efficient transport of electrons because of their high surface-to-volume ratio and tunable electron transport properties due to quantum confinement effect [26]. In comparison to the 2D structure, they avoid reduction in signal intensities as well as provide sensing modality for label-free and direct electrical readout when the nanostructure is used as a semiconducting channel for a chemiresistor [27]. Such label-free and direct detection is highly desirable for rapid and real-time monitoring of receptor–ligand interaction with a modified receptor nanostructure, particularly when the receptor is a biomolecule such as antibody, DNA, and protein. This is critical for clinical diagnosis and biowarfare agent's detection applications. Table 5.1 provides a summary of the various thin film structures used for biosensor application.

TABLE 5.1 Summary of Overall Performances of Thin Film-Based Biosensors

SL No.	Analyte	Enzyme	Immobilization Method	Analytical Method	Nanostructure	Ref.
1.	Glucose	Gox		Quenching	Al doped ZnO TF	[28]
2.	Glucose	Gox	Electrocatalytic	Potentiometric	Cu/Ni TF alloy	[29]
3.	Glucose	Gox	Electrochemical	Amperometric	SnO2 TF	[30]
4.	Uric acid	Uricase	Colorimetric		Ni TF	[31]
5.	Uric acid	Uricase	Electrochemical	Potentiometric	ZnO TF/CuO	[32]
6.	H_2O_2	Hb GCE	Entrapped	Amperometric	ZnO TF	[33]
7.	H_2O_2	HRP	Electrocatalytic	Amperometric	Cu doped ZnO TF	[34]
8.	Glucose	Gox			CuO TF	[35]
9.	Uric acid	Uricase	Adsorption	Amperometric	NiO TF	[36]

5.3.3 Nanowires

Among other 1D nanostructure, nanowire (NW)-based sensors have the advantages of (1) sensitivity, (2) spatial resolution, (3) rapid response associated with individual NW, (4) ease of fabrication methods allowing large scale production, (5) dimension which can be compared to the space charge region, (6) high crystallinity obtained due to superior stability, and (7) direct and real-time transducers at room temperature [36]. In the case of metallic NWs, the electrons conduct while transporting through the wire experiencing multiple diffusive scattering. On reducing the length, the electron transport becomes ballistic, and the resistance affects the adsorbates. So, the conduction changes depending on the molecular adsorption on the surface of the metal NWs. Such metal nanowires can be used for virus detection such as copper nanowires for the central channel of tobacco mosaic virus particles [37, 38] and Au/Ag striped nanowires as multiplexed immunoassay platforms for pathogen detection are shown [39]. In the case of the semiconducting NWs, the change in the carrier concentration could modify the device performance [38]. The surface of semiconductor NWs can be covered with small amounts of catalyst NPs (noble metals, Pt, or Pd), such that these catalysts accelerate the dissociation of the molecular test gas into its more reactive atomic form, which will accordingly result in an enhanced surface reaction during gas sensing. Moreover, treating the surface of the NWs with oxidizing (O_2) and reducing plasma (NH_3) causes a respective modulation (decrease and increase) of the charge carrier concentration [39]. Also, the grain size, structure, surface roughness, and configuration of the sensor could affect the sensor performance. Electrical conductivity strongly depends on the surface state produced by molecular adsorption that results in space-charge layer changes and band modulation. Thus, the change in the conductance in the case of the NW is obtained from the electron hopping between the adsorbed molecule and the NW, the chemical potential, and the change of atomistic configurations of the NW near the adsorption site. Thus, Table 5.2 provides a summary of various 1D nanowires-based biosensors. Moreover, Silicon NWs have a distinct contribution at the valence band maximum or at the conduction band minimum coming from atoms at distinct surfaces [40]. Silicon atoms are bonded to the core atoms in such a particular way showing unique capabilities in protein, virus, and DNA detection for diagnosis of various diseases and drug discovery.

TABLE 5.2 Summary of Overall Performances of Nanowire-Based Biosensors

SL No.	Analyte	Enzyme	Immobilization Method	Analytical Method	Nanostructure	Ref.
1.	Glucose	Gox	Adsorption	Amperometric	ZnO NW/ GrapheneNP	[42]
2.	Glucose	Gox	Electrostatic	Potentiometric	Au/Ni/Au NW	[43]
3.	Glucose	Gox	Electrochemical	Amperometric	TiO_2 NW/GOx	[44]
4.	Pesticides	AChE	Electrochemical	Amperometric	Pd-Cu NW	[45]
5.	Uric acid	Uricase	Crosslinking	Amperometric	ZnO NW	[46]
6.	Uric acid	Uricase	Crosslinking	Amperometric	Cu_2O/Cu@C core shell NW	[47]
7.	H_2O_2	HRP	Electrocatalytic	Amperometric	Cu_2O NW	[48]
8.	H_2O_2	HRP	Electrocatalytic	Amperometric	Cu NP/Si NW	[49]

5.3.4 Carbon Nanotubes (CNT)

Carbon nanotubes (CNTs) are cylindrical nanostructures comprising "rolled-up" sheets of a single layer of carbon atoms. They can be classified depending on their structure as either single-walled or multi-walled carbon nanotubes. The SWCNT is formed by rolling a single graphite sheet into a tube with diameters between 0.4 and 2.5 nm whereas MWCNTs are composed of multiply-nested graphene sheets, like rings of a tree trunk, with variable diameters up to 100 nm and with length from a few nanometers to several micrometers. CNTs offer a large surface–volume ratio which enables immobilization of much larger quantities of biomolecules to develop novel probes for a variety of biomolecules. CNTs can be used for biosensors as (1) they capture specific platforms, (2) transduction of analyte-induced electrical signals using probes, and (3) novel mechanism of in vivo probing due to the ability of CNTs to cross a biological membrane [50]. In the case of enzymatic electrochemical biosensors, the electron transfers between the redox-active center of enzyme and the electrode are the critical point for the optimal performance. They are also low-cost, easy fabrication processes to interact with ions metabolites, and biomarkers for different applications with enzymes (glucose, cholesterol, nitric oxide, etc.) and proteins (biofuel cell, ethanol, glucose, nitrite, and oxygen) [51]. CNT-based biosensors have also been developed for accurate sensing of cholesterol using electrodes functionalized with holesterolesterase, peroxidase, and oxidase [52, 53]. In the case of electrochemical immunosensors, CNTs are deposited with antigens with biomolecules to amplify the detected signal. DNA-based electrochemical sensors were also investigated for the detection of genome

TABLE 5.3 Summary of Overall Performances of CNT-Based Biosensors

SL No.	Analyte	Enzyme	Immobilization Method	Analytical Method	Nanostructure	Ref.
1.	Glucose	Gox	Adsorption	Amperometric	ZnO/CNTs	[56]
2.	Glucose	Gox	Electrostatic	Amperometric	ZnO/CNTs	[57]
3.	H_2O_2	HRP	Adsorption	Amperometric	TiO_2/Au/CNT	[58]
4.	Phenol	Tyr	Entrapment	Amperometric	TiO_2/PVA/CNT	[59]
5.	Phenol	Tyr	Crosslinking	Amperometric	TiO_2/CNT	[60]
6.	Uric acid	Uricase	Electrostatic	Amperometric	ZnO/CNT	[61]
7.	L-lactate	LOD	Electrostatic	Amperometric	ZnO/CNT	[62]
8.	Catechol	Tyr	Covalent	Amperometric	Fe_3O_4/CNT	[63]
9.	Glucose	Gox	Electrocatalytic	Amperometric	Cu_2O/MWCNTs nanocomposites	[64]
10.	Glucose	Gox	Electrochemical	Amperometric	CuO/RGO/CNT	[65]

mutation and infectious diseases [54]. The optical biosensors are designed for detecting, investigating, and quantifying biological processes in vitro and in vivo. Such biosensors can be classified based on their specific optical transduction mechanism, for example, surface plasmon resonance, fluorescence, absorbance/reflectance, etc. [55]. They can also be engineered to display changes in fluorescent emission signals upon exposure to the intended analyte. Table 5.3 provides a list of CNT-based structures used for biosensor applications.

5.3.5 Nanoparticles

They are defined as isolable elements with dimensions included between 1 and 50 nm and due to their size nanoparticles (NPs) have physical, electronic, and chemical properties quite different from bulk materials [66]. Various types of NPs, including metal NPs, oxide NPs, semiconductor NPs, and composite NPs, have been widely used for developing electrochemical, optical, and piezoelectric biosensors [67–69]. However, NPs have demonstrated augmented sensitivity against certain analytes and have therefore been extensively used for biosensing purposes as shown in Table 5.4. Various reports on metallic NP using gold and silver and semiconducting NP such as CdS NP have integrated for efficient and sensitive biosensing [70–76]. Also, NPs have as important feature of high specific surface area and high surface-free energy and as a result they are able to adsorb more promoting the immobilization step without any denaturation process and loss of bioactivity used as biomarkers in

TABLE 5.4 Summary of Overall Performances of Nanoparticle-Based Biosensors

SL No.	Analyte	Enzyme	Immobilization Method	Analytical Method	NP Decorated	Ref.
1.	Glucose	Gox	Electrostatic	Amperometric	ZnO	[85]
2.	Glucose	Gox	Covalent	Amperometric	TiO_2	[86]
3.	Glucose	Gox	Adsorption	Amperometric	Fe_3O_4	[87]
4.	Glucose	Gox	Crosslinking	Amperometric	ZnO	[88]
5.	Cholesterol	ChOx	Adsorption	Amperometric	ZnO	[89]
6.	Cholesterol	ChOx	Electrostatic	Amperometric	ZnO	[90]
7.	Cholesterol	ChOx	Electrostatic	Amperometric	ZnO	[91]
8.	H_2O_2	HRP	Electrostatic	Amperometric	ZnO/Au	[92]
9.	H_2O_2	HRP	Entrapment	Amperometric	TiO_2	[93]
10.	Phenol	Tyr	Entrapment	Amperometric	TiO_2	[94]
11.	Urea	Urease	Electrostatic	Amperometric	ZnO	[95]
12.	Uric acid	Uricase	Adsorption	Amperometric	ZnO	[96]
13.	Pesticide	AChE	Electrostatic	Amperometric	ZnO	[97]

the case of biomolecules. They also have high chemical stability with fast electron transport capability. Various other features of NP include enhanced catalytic property using various metal NP such as Au, Pt, Ni, and Cu [77, 78]. Use of such NP could provide excellent properties in the case of biomolecules with innovative biosensor system. Among different NP, most used metal NP is gold (Au) NP due to their biocompatibility, unique opto-electronic properties, and their relatively simple fabrication and modification techniques. Moreover, gold promotes fast and direct electron transfer between a wide range of electroactive species due to its excellent conducting properties along with light-scattering properties. In gold nanoparticle-assisted biosensing, the collective oscillation of plasmons is in response to incident electromagnetic (EM) waves where the phenomenon of surface plasmon resonance takes place. Lin et al. who used the SPR technique demonstrated an optical fiber-based biosensor that is sensitive to organophosphorus pesticides [79]. The presence of pesticides would lead to a change in the attenuation of the incident light. Au NPs can immobilize a far greater number of protein molecules, which therefore allows the realization of more sensitive biosensor devices due to their high surface area. Andreescu et al. utilized immobilized periplasmic glucose receptors on Au NPs and measured their electrical properties as a function of glucose concentration to develop a novel glucose sensor [80]. Au NPs have also been effectively utilized to reduce the overpotentials of crucial electrochemical reactions as well as engineer the reversibility of certain redox reactions [81–83]. Tuener et al. demonstrated that 1.4-nm-sized Au NPs provided catalyst-like features for the oxidation of styrene by dioxygen [84]. Moreover, reports on metal-oxide-based NP (MONP) were also used as the electrode where they provide ideal immobilization matrices as well as transduction platform and/or mediators to improve the sensitivity of the device. As such these MONP devices were used in the development of immune, enzymes, DNA, and enzymatic biosensors.

5.4 APPLICATIONS OF BIOSENSORS

Due to their higher sensitivity, good stability, and better selectivity, biosensors are very useful in various fields like industry, medicine, and various ecofriendly applications. The detailed study of various applications of these biosensors in various fields is given in Figure 5.3.

5.4.1 Industrial Applications

5.4.1.1 In Food Processing, Beverages Monitoring, Quality, and Safety

Food is the essential thing for all human beings. It is important to produce good food by controlling and monitoring the unwanted things in the food industry. So, biosensors are used in the food processing industry due to their necessity in the production of good food for the enormous population. Quality and safety, maintenance of food products, and processing are the important parameters in the food processing industry. For that, we need

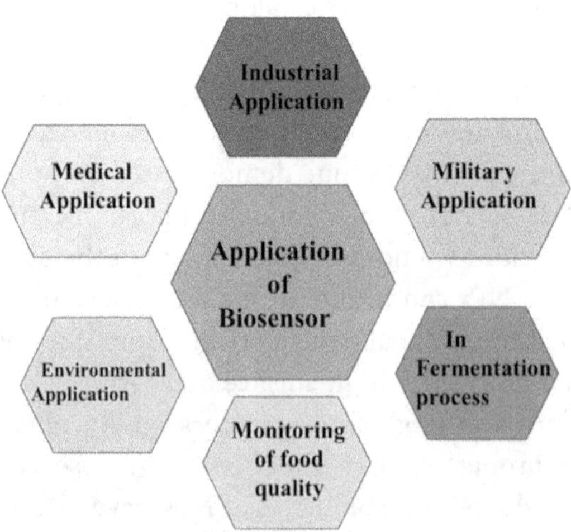

FIGURE 5.3 Different applications of biosensors.

TABLE 5.5 Biosensors Used in Different Industries

Sl. No	Industry	Sensing for	Sensor Type	Used Material/ Technique	Ref.
1.	Dairy industry	Organophosphate pesticides in milk	Enzymatic biosensors	Screen-printed carbon electrode	[98]
2.	Beverage industry	Ageing of beer	Enzymatic biosensors	Cobalt phthalocyanine	[99]
3.	Sweet food industry	Electrophysiological activities of the taste epithelium (artificial sweeteners)	Electrochemical techniques	Spatiotemporal techniques	[100]
4.	Vegetables	Pathogens in food	Potentiometric alternating bio sensing	Variation in pH caused by ammonia	[101]
5.	Fermentation industry	Monitoring of Saccharification (glucose, lactate, lysine, ethanol)	Biochemical sensor	Traditional Fehiing's method	[102]

some chemical experiments as well as some sensors to monitor. But using chemical experiments and spectroscopy are the traditional techniques that have less preference due to human fatigue, are expensive, and are more time-consuming. Thus, the use of biosensors in response to the demand for simple, real-time, selective, and inexpensive techniques is highly preferable. Table 5.5 shows the various biosensors used in various industries for detecting different parameters.

5.4.2 Medical Applications

5.4.2.1 In Medical Field

The applications of biosensors in the field of medical science are growing rapidly due to their enormous requirement. For the diagnosis of diabetes mellitus, glucose biosensors are widely used in clinical applications which require good control over blood glucose levels [103]. Some other biosensors are being used pervasively in the medical field to diagnose infectious diseases. For diagnosis of urinary tract infection (UTI) along with pathogen recognition and anti-microbial susceptibility, biosensor technology is under process. The major problem faced today is heart failure with about one million people suffering from it. Hafnium oxide (HfO_2) biosensor has worked greatly for antibody deposition with the detection of a human antigen by electrochemical impedance spectroscopy techniques for the detection of cardiovascular diseases like immunoaffinity column assay, fluorometric, and enzyme-linked immunosorbent assay. For that, the biosensors are designed to generate electric measurements by the biochemical molecular identification [104, 105]. Some of the other biosensors used in different medical applications are as follows: (i) Cardiac markers with quantitative measurement in undiluted serum. (ii) To control endothelin induced due to cardiac hypertrophy, the microfluidic impedance assay is used. (iii) The immune sensor array used for clinical immune phenotyping of acute leukemias. (iv) In dental diseases, to sense the effect of oxazaborolidines on immobilized fructosyltransferase. (v) The resonance energy transfer of Histone deacetylase (HDAC) inhibitor assay. (vi) For quick detection of multiple cancer markers, the biochip is used. (vii) Diamond microneedle electrodes are used for accurate and neurochemical detection.

5.4.2.2 Fluorescent Biosensors

In general, these fluorescent biosensors are imaging agents that are used to find cancer cells and drug discovery where they enable insights into the role and regulation of enzymes at the cellular level. For that, genetically encoded FRET and GFP-based biosensors play an important role in sensing mechanism. In the fluorescent biosensors, there are one or more fluorescent probes placed (enzymatically, chemically, or genetically) through a receptor. The function of the receptor is to identify a specific analyte or target, thereby converting a fluorescent signal that can be readily detected and measured [106]. These fluorescent biosensors have high sensitivity to ions, metabolites, and protein biomarkers. These fluorescent biosensors are also used to identify the existence, activity, or status of the target (serum, cell extracts) in complex solutions. They are used in probing gene expression, protein localization, and conformation in fields such as cell cycle, signal transduction, apoptosis, and transcription. By using these sensors, we can get the indication of arthritis, inflammatory diseases, cardiovascular and neurodegenerative diseases, viral infection, cancer, and metastasis. These fluorescent biosensors are used in drug discovery programs leading to the recognition of drugs with high throughput, high content screening approaches, also in post-screening analysis of hits and optimization of leads. For preclinical evaluation, these sensors are considered potent tools for clinically validating the therapeutic potential, biodistribution, and pharmacokinetics of candidate drugs [107]. Fluorescent biosensors are effectively employed for the early detection of biomarkers in molecular and clinical diagnostics, for monitoring disease progression and response to treatment/therapeutics, for intravital imaging and image-guided surgery [108].

5.4.3 Military Applications

The biological attack is the most dangerous threat that destroys the physical and economic wealth of the particular territory. Biosensors are very useful for military purposes at times of these biological attacks. The main motive of such biosensors is to identify organisms posing a threat sensitively and selectively. These organisms in virtually real-time are called biowarfare agents (BWAs) and are generally bacterium (vegetative and spores), toxins, and viruses. To recognize the chemical markers of BWAs, such biosensors have been using molecular techniques with several attempts by using nucleic acid-based sensing or antibody-based detection. The nucleic acid-based sensing systems are well-sensitive and more preferred than antibody-based detection methods because they provide gene-based specificity, and they do not need any amplification steps to attain detection sensitivity to the preferred levels. A novel leaky surface acoustic wave peptide nucleic acid biosensor with double two-port resonators can detect human papillomavirus HPV (double-stranded DNA virus: HPV 16 and 18) which is related to invasive cervical cancer. This probe directly detects HPV genomic DNA without polymerase chain reaction amplification and can also bind to the target DNA sequences with a lot of efficacy and precision.

5.4.4 Biosensors for Environment

5.4.4.1 Biosensing Technology for Sustainable Food Safety

The term food quality means the recognition of the food in its appearance, taste, smell, nutritional value, freshness, flavor, texture, and chemicals [109]. In food quality and safety measurement, smart monitoring of nutrients and fast screening of biological and chemical contaminants are very important parameters. For good health, efforts are being made to develop control systems to monitor different parameters in the generated food. Glucose monitoring becomes indispensable as during storage the food content and composition may get altered. German studied the electrochemistry of glucose oxidase immobilized on a graphite rod, altered by gold nanoparticles (AuNPs), which improved its sensitivity [110]. Glutaminase-based microfluidic biosensor chip with a flow-injection analysis for electrochemical detection has been used for detection in the fermentation process [111]. Biosensors are being employed to perceive general toxicity and specific toxic metals, due to their capability to react with only the hazardous fractions of metal ions [112]. In general, pesticides pose grave threats to the environment. Immunosensors have produced their merit as a sensitive, good detector of common pesticides organophosphates, and carbamic insecticide species, high-speed agrifood, and environmental monitoring. Biosensors based on AChE and butyryl cholinesterase have been designed for the detection of aldicarb, carbaryl, paraoxon, chlorpyrifos methyl, etc. Oxon utilizing screen-printed electrodes biosensor is used to detect pesticides in wine and orange juice [113]. In some food elements, Arsenic can be measured with the help of bacteria-based bioassays.

5.4.4.2 Biosensors in Plant Biology

The advancements in plant science are based on DNA sequencing and molecular imaging for developing different technologies. Cellular and subcellular localization, and measure of ion and metabolite levels are the general methods of mass spectroscopy for gauging insights into unprecedented precision. However, it is very difficult to find the key information regarding location and dynamics of enzyme substrates, receptors, and transporters. However, by using the biosensors this information can easily and successfully be tapped. We need to devise tactics to visualize the actual process to measure a dynamic process under physiological conditions. For instance, the conversion of one metabolite takes place into another or triggers signaling events. This visualization can be done by dynamically responding to biosensors. Roger Tsien's lab was the first to develop protein prototype sensors to measure caspase activity and control levels of calcium in live cells [114]. These sensors were based on FRET between two spectral variants of GFP [115]. In vivo application of biosensors involves high temporal resolution imaging of calcium oscillations using chameleon sensors. Plant biosensors can be utilized to identify missing components pertinent to the metabolism, regulation, or transport of the analyte. For sucrose, FRET sensors are designed and are responsible for the identification of proteins and perform a transport step in phloem loading sucrose efflux from the mesophyll. Fluorometric-based assays with FRET sugar sensors successfully recognize the sugar transporters that can function immediately after exposure of starved yeast cells to glucose [116].

5.4.4.3 In Metabolic Engineering

The lack of sustainability of petroleum-derived products is gradually exhorting the need for the establishment of microbial cell industries for the synthesis of chemicals for environmental concerns. For increasing bioeconomy, the researchers view metabolic engineering as the enabling technology [117]. They have also envisioned that a substantial fraction of fuels, commodity chemicals, and pharmaceuticals will be produced from renewable feedstocks by exploiting microorganisms rather than relying on petroleum refining or extraction from plants. The high capacity for diversity generation also requires efficient screening methods to select the individuals carrying the desired phenotype. The earlier methods were spectroscopy-based enzymatic assay analytics; however, they had limited throughput. To circumvent this obstacle genetically encoded biosensors that enable in vivo monitoring of cellular metabolism were developed which offered the potential for high throughput screening and selection using fluorescence-activated cell sorting (FACS) and cell survival, respectively. Transcription factors are natural sensory proteins evolved to regulate gene expression in response to changes in the environment for high throughput screening. It is accomplished by hacking into the host transcription system and employing a synthetic condition-specific promoter to drive the expression of a reporter gene. These exhibit poor orthogonality and background noise [118]. The third class of biosensors comprises riboswitches: the regulatory domain of an mRNA that can selectively bind to a ligand and thereupon change its own structure, consequently regulating transcription of its encoded protein. As opposed to TF-based biosensors, they are comparatively faster as the RNA has already been transcribed, also they do not rely on protein–protein or protein–metabolite interactions. In recent decades, ribosomes have been extensively engineered in bacterial systems [119].

5.5 CONCLUSION AND FUTURE CHALLENGES

The application of nanostructured metal oxides-based biosensors has been studied in this chapter which provides a review of present research status and applications. Also, different structures like thin films, one-dimensional, nanowires, CNT, and nanoparticles-based biosensors were studied in detail. These nanostructured-based biosensors represent a powerful detection platform for electrochemical enzymatic biosensors. It is a crucial requirement of consumers to design simple, affordable, reliable, and portable electrochemical biosensors that provide excellent features adding high sensitivity, selectivity, fast response, and low detection limit with minimum interferences. Electrochemical biosensors could be useful for diagnosing and monitoring infectious diseases, monitoring the pharm kinetics of drugs, detecting cancer, and disease biomarkers, and analyzing breath, urine, and blood. The highly sensitive detection protocol for the fabrication of biosensors using metal oxides is recommended as a future research element which is likely to lead to a new generation of nanostructured-based biosensors.

REFERENCES

1. Liu A. Towards development of chemosensors and biosensors with metal-oxide-based nanowires or nanotubes. *Biosensors and Bioelectronics.* 2008;24(2):167–177.
2. Preda G, Bizerea O, Vlad-Oros B. Sol-gel technology in enzymatic electrochemical biosensors for clinical analysis. *Biosensors for Health, Environment and Biosecurity.* InTech. 2011. doi:10.5772/19622. ISBN: 978-953-307-443-6
3. Mehrotra P. Biosensors and their applications–A review. *Journal of Oral Biology and Craniofacial Research.* 2016;6(2):153–159.
4. Batool R, Rhouati A, Nawaz MH, Hayat A, Marty JL. A review of the construction of nano-hybrids for electrochemical biosensing of glucose. *Biosensors.* 2019;9(1):46.
5. Dakshayini BS, Reddy KR, Mishra A, Shetti NP, Malode SJ, Basu S, Naveen S, Raghu AV. Role of conducting polymer and metal oxide-based hybrids for applications in amperometric sensors and biosensors. *Microchemical Journal.* 2019;147:7–24.
6. Ali J, Najeeb J, Asim Ali M, Aslam MF, Raza AJ. Biosensors: their fundamentals, designs, types and most recent impactful applications: a review. *J. Biosens. Bioelectron.* 2017;8(1):1–9.
7. Demirkan B, Bozkurt S, Şavk A, Cellat K, Gülbağca F, Nas MS, Alma MH, Sen F. Composites of bimetallic platinum-cobalt alloy nanoparticles and reduced graphene oxide for electrochemical determination of ascorbic acid, dopamine, and uric acid. *Scientific Reports.* 2019;9(1):1–9.
8. Shi X, Gu W, Li B, Chen N, Zhao K, Xian Y. Enzymatic biosensors based on the use of metal oxide nanoparticles. *Microchimica Acta.* 2014;181(1–2):1–22.
9. LaFleur L, Yager P. Medical biosensors. In *Biomaterials Science.* 2013; 996–1006. Academic Press.
10. Solanki PR, Kaushik A, Agrawal VV, Malhotra BD. Nanostructured metal oxide-based biosensors. *NPG Asia Materials.* 2011;3(1):17–24.
11. Rahman Md, Ahammad AJ, Jin J-H, Ahn SJ, Lee J-J. A comprehensive review of glucose biosensors based on nanostructured metal-oxides. *Sensors.* 2010;10(5):4855–4886.

12. Ansari AA, Alhoshan M, Alsalhi MS, Aldwayy AS. Nanostructured metal oxides based enzymatic electrochemical biosensors. *Biosensors*. InTech. 2010. doi:10.5772/7201. ISBN: 978-953-7619-99-2
13. Waryo T, Kotzian P, Begić S, Bradizlova P, Beyene N, Baker P, Kgarebe B, et al. Amperometric hydrogen peroxide sensors with multivalent metal oxide-modified electrodes for biomedical analysis. In *13th International Conference on Biomedical Engineering*, 2009; 829–833. Springer, Berlin, Heidelberg.
14. Malhotra BD, Das M, Solanki PR. Opportunities in nano-structured metal oxides based biosensors. *Journal of Physics: Conference Series*. 2012;358(1):012007.
15. Tereshchenko AV, Smyntyna VA, Konup IP, Geveliuk SA, Starodub MF. Metal oxide based biosensors for the detection of dangerous biological compounds. *Nanomaterials for Security*, 2016;281–288. Springer, Dordrecht.
16. Agah A, Hassibi A, Plummer JD, Griffin PB. Design requirements for integrated biosensor arrays. European Conference on Biomedical Optics. Munich, Germany, 12 June 2005.
17. Patolsky F, Zheng G, Lieber CM. Nanowire sensors for medicine and the life sciences. *Nanomedicine*. 2006;1(1):51–65.
18. Carrara S, Ghoreishizadeh S, Olivo J, et al. Fully integrated biochip platforms for advanced healthcare. *Sensors Basel*. 2012;12(8):11013–11060.
19. Gooding JJ, Lai LM, Goon IY. Nanostructured electrodes with unique properties for biological and other applications. *Chemically Modified Electrodes*. 2009;11:1–56.
20. Shanmugam NR, Muthukumar S, Prasad S. A review on ZnO-based electrical biosensors for cardiac biomarker detection. *Future Science OA*. 2017;3(4):FSO196.
21. Aishwaryadev Banerjee HK, Shakirul HK, Broadbent S, Bulbul A, Kim KH, Looper R, Mastrangelo CH. Molecular bridge mediated ultra-low-power gas sensing. *arXiv.org*, 2019, doi: https://arxiv.org/abs/1911.05965.
22. Banerjee et al. Molecular bridge mediated ultra-low-power gas sensing. Nov. 2019, [Online]. Available: http://arxiv.org/abs/1911.05965.
23. Sokolov N, Tee BC-K, Bettinger CJ, Tok JB-H, Bao Z, Chemical and engineering approaches to enable organic field-effect transistors for electronic skin applications. *Acc Chem Res*. 2012;45(3):361–371, doi: 10.1021/ar2001233.
24. Böcking T, James M, Coster H-GL, Chilcott TC, Barrow KD. Structural characterization of organic multilayers on silicon (111) formed by immobilization of molecular films on functionalized Si–C linked monolayers. *Langmuir*. 2004;20(21):9227–9235.

25. Marmont P, Battaglini N, Lang P, Horowitz G, Hwang J, Kahn A, Amato C, Calas P. Improving charge injection in organic thin-film transistors with thiol-based self-assembled monolayers. *Organic Electronics.* 2008;9(4):419–424.
26. Wanekaya AK, Chen W, Myung NV, Mulchandani A. Nanowire-based electrochemical biosensors. *Electroanalysis: An International Journal Devoted to Fundamental and Practical Aspects of Electroanalysis.* 2006;18(6):533–550.
27. Janata J, Josowicz M. Conducting polymers in electronic chemical sensors. *Nature Materials.* 2003;2(1):19–24.
28. Ghosh J, Ghosh R, Giri PK. Tuning the visible photoluminescence in Al doped ZnO thin film and its application in label-free glucose detection. *Sensors and Actuators B: Chemical.* 2018;254:681–689.
29. Pötzelberger I, Mardare AI, Hassel AW. Non-enzymatic glucose sensing on copper-nickel thin film alloy. *Applied Surface Science.* 2017;417:48–53.
30. Haider AJ, Mohammed AJ, Shaker SS, Yahya KZ, Haider MJ. Sensing characteristics of nanostructured SnO_2 thin films as glucose sensor. *Energy Procedia.* 2017;119:473–481.
31. Tripathi A, Harris KD, Elias AL. Peroxidase-like behavior of Ni thin films deposited by glancing angle deposition for enzyme-free uric acid sensing. *ACS Omega.* 2020:5(16):9123–9130.
32. Jindal K, Tomar M, Gupta V. A novel low-powered uric acid biosensor based on arrayed pn junction heterostructures of ZnO thin film and CuO microclusters. *Sensors and Actuators B: Chemical.* 2017;253:566–575.
33. Ma W, Song W, Tian DB. ZnO-MWCNTs/Nafion inorganic-organic composite film: Preparation and application in bioelectrochemistry of haemoglobin. *Chin Chem Lett.* 2009;20:358–361.
34. Asikuzun E, Ozturk O, Arda L, Terzioglu C. Preparation, growth and characterization of nonvacuum Cu-doped ZnO thin films. *Journal of Molecular Structure.* 2018;1165:1–7.
35. Lillo-Ramiro J, Guerrero-Villalba JM, Mota-González ML, Aguirre-Tostado FS, Gutiérrez-Heredia G, Mejía-Silva I, Carrillo-Castillo A. Optical and microstructural characteristics of CuO thin films by sol gel process and introducing in non-enzymatic glucose biosensor applications. *Optik.* 2021;229:166238.
36. Arora K, Tomar M, Gupta V. Highly sensitive and selective uric acid biosensor based on RF sputtered NiO thin film. *Biosensors and Bioelectronics.* 2011;30(1):333–336.
37. Balci S, Bittner AM, Hahn K, Scheu C, Knez M, Kadri A, Wege C, Jeske H, Kern K. Detection in real time which are exploited in vivo diagnostics. *Electrochim Acta.* 2006;51:6251–6257.

38. Balci S, Noda K, Bittner A, Kadri A, Wege C, Jeske H, Kern K. Self-assembly of metal–virus nanodumbbells. *Angew Chem Intl Ed*. 2007;46: 3149–3151.
39. Tok JB-H, Chuang FYS, Kao MC, Rose KA, Pannu SS, Sha MY, Chakarova G, Penn SG, Dougherty GM. Metallic striped nanowires as multiplexed immunoassay platforms for pathogen detection. *Angew Chem Int Ed*. 2006;45:6900–6904.
40. Law JBK, Thong JTL. Improving the NH3 gas sensitivity of ZnO nanowire sensors by reducing the carrier concentration. *Nanotechnology*. 2008;19:205502.
41. Cui Y, Wei Q, Park H, Lieber CM. Nanowire nanosensors for highly sensitive and selective detection of biological and chemical species. *Science*. 2001;293:1289–1292.
42. Rafiee Z, Mosahebfard A, Sheikhi MH. High-performance ZnO nanowires-based glucose biosensor modified by graphene nanoplates. *Materials Science in Semiconductor Processing*. 2020;115:105116.
43. Asif MH, Razaq A, Akbar N, Danielsson B, Sultana I. Facile synthesis of multisegment Au/Ni/Au nanowire for high performance electrochemical glucose sensor. *Materials Research Express*. 2019;6(9):095028.
44. Yang W, Wang X, Hao W, Wu Q, Peng J, Tu J, Cao Y. 3D hollow-out TiO_2 nanowire cluster/GOx as an ultrasensitive photoelectrochemical glucose biosensor. *Journal of Materials Chemistry B*. 2020;8(11):2363–2370.
45. Song D, Li Y, Lu X, Sun M, Liu H, Yu G, Gao F. Palladium-copper nanowires-based biosensor for the ultrasensitive detection of organophosphate pesticides. *Analytica Chimica Acta*. 2017;982:168–175.
46. Luo J, Cui J, WangY, Yu D, Qin Y, Zheng H, Shu X, Tan HH, Zhang Y, Wu Y. Metal-organic framework-derived porous Cu_2O/Cu@ C core-shell nanowires and their application in uric acid biosensor. *Applied Surface Science* 2020;506:144662.
47. Liu X, Lin P, Yan X, Kang Z, Zhao Y, Lei Y, Li C, Du H, Zhang Y. Enzyme-coated single ZnO nanowire FET biosensor for detection of uric acid. *Sensors and Actuators B: Chemical*. 2013;176:22–27.
48. Lu W, Sun Y, Dai H, Ni P, Jiang S, Wang Y, Li Z, Li Z. Direct growth of pod-like Cu2O nanowire arrays on copper foam: highly sensitive and efficient nonenzymatic glucose and H2O2 biosensor. *Sensors and Actuators B: Chemical*. 2016;231:860–866.
49. Huang J, Zhu Y, Zhong H, Yang X, Li C. Dispersed CuO nanoparticles on a silicon nanowire for improved performance of nonenzymatic H_2O_2 detection. *ACS Applied Materials & Interfaces*. 2014;6(10):7055–7062.
50. Banerjee A, Maity S, Mastrangelo CH. Nanotechnology for biosensors: A review. *arXiv preprint arXiv:2101.02430* (2021).

51. Nagaraju K, Reddy R, Reddy N. A review on protein functionalized carbon nanotubes. *Journal of Applied Biomaterials & Functional Materials.* 2015;13(4):301–312.
52. Li G, Liao JM, Hu GQ, Ma NZ, Wu PJ, Study of carbon nanotube modified biosensor for monitoring total cholesterol in blood. *Biosens Bioelectron.* 2005;20(10):2140–2144, doi: 10.1016/j.bios.2004.09.005.
53. Santos RM, Rodrigues MS, Laranjinha J, Barbosa RM. Biomimetic sensor based on hemin/carbon nanotubes/chitosan modified microelectrode for nitric oxide measurement in the brain. *Biosens Bioelectron.* 2013;44:152–159, doi: 10.1016/j.bios.2013.01.015.
54. Wang J. Survey and summary: from DNA biosensors to gene chips. *Nucleic Acids Research.* 2000;28(16):3011–3016.
55. Yang R, Jin J, Chen Y, Shao N, Kang H, Xiao Z, Tang Z, Wu Y, Zhu Z, Tan W. Carbon nanotube-quenched fluorescent oligonucleotides: probes that fluoresce upon hybridization. *J Am Chem Soc.* 2008;130:8351–8358.
56. Hu FX, Chen SH, Wang CY, Yuan R, Chai YQ, Xiang Y, Wang C. ZnO nanoparticle and multiwalled carbon nanotubes for glucose oxidase direct electron transfer and electrocatalytic activity investigation. *J Mol Catal B.* 2011;72(3–4):298–304.
57. Wang YT, Yu L, Zhu ZQ, Zhang J, Zhu JZ, Fan CH. Improved enzyme immobilization for enhanced bioelectrocatalytic activity of glucose sensor. *Sensors Actuators B.* 2009;136(2):332–337.
58. Zhong HA, Yuan R, Chai YQ, Li WJ, Zhang Y, Wang CY. Amperometric biosensor for hydrogen peroxide based on horseradish peroxidase onto gold nanowires and TiO_2 nanoparticles. *Bioprocess Biosyst Eng.* 2011;34(8):923–930.
59. Tan SW, Tan XC, Xu J, Zhao DD, Zhang JL, Liu L. A novel hydrogen peroxide biosensor based on sol-gel poly (vinyl alcohol) (PVA)/(titanium dioxide)TiO_2 hybrid material. *Anal Methods.* 2011;3(1):110–115.
60. Lee YJ, Lyn YK, Choi HN, Lee WY. Amperometric tyrosinase biosensor based on carbon nanotube-titania-Nafion composite film. *Electroanalysis.* 2007;19(10):1048–1054.
61. Wang YT, Yu L, Zhu ZQ, Zhang J, Zhu JZ. Novel uric acid sensor based on enzyme electrode modified by ZnO nanoparticles and multiwall carbon nanotubes. *Anal Lett.* 2009;42(5):775–789.
62. Devi R, Yadav S, Pundir CS. Amperometric determination of xanthine in fish meat by zinc oxide nanoparticle/chitosan/multiwalled carbon nanotube/polyaniline composite film bound xanthine oxidase. *Analyst.* 2012;137(3):754–759.
63. Peacuterez-Loacutepez B, Merkoccedili A. Magnetic nanoparticles modified with carbon nanotubes for electrocatalytic magnetoswitchable biosensing applications. *Adv Funct Mater.* 2011;21(2):58–6363.

64. Zhang X, Wang G, Zhang W, Wei Y, Fang B. Fixure-reduce method for the synthesis of Cu_2O/MWCNTs nanocomposites and its application as enzyme-free glucose sensor. *Biosens Bioelectron.* 2009;24:3395–3398.
65. Lee C, Lee SH, Cho M, Lee Y. Nonenzymatic amperometric glucose sensor based on a composite prepared from CuO, reduced graphene oxide, and carbon nanotube. *Microchimica Acta.* 2016;183(12):3285–3292.
66. Bönnemann H, Richards RM. Nanoscopic metal particles–synthetic methods and potential applications. *European Journal of Inorganic Chemistry.* 2001;10:2455–2480.
67. Raj CR, Abdelrahman AI, Ohsaka T. Gold nanoparticle-assisted electroreduction of oxygen. *Electrochemistry Communications.* 2005;7(9): 888–893.
68. Lin H, Musick M-D, Nicewarner SR, Salinas FG, Benkovic SJ, Natan MJ, Keating CD. Colloidal Au-enhanced surface plasmon resonance for ultrasensitive detection of DNA hybridization. *Journal of the American Chemical Society.* 2000;122(38):9071–9077.
69. Zhou J, Gan N, Li T, Zhou H, Li X, Cao Y, Wang L, Sang W, Hu F. Ultratrace detection of C-reactive protein by a piezoelectric immunosensor based on Fe3O4@ SiO2 magnetic capture nanoprobes and HRP-antibody co-immobilized nano gold as signal tags. *Sensors and Actuators B: Chemical.* 2013;178:494–500.
70. Xiao Y, Patolsky F, Katz E, Hainfeld JF, Willner I. "Plugging into Enzymes": Nanowiring of redox enzymes by a gold nanoparticle. *Science.* 2003:299(5614):1877–1881, doi: 10.1126/science.1080664.
71. Schierhorn M, Lee SJ, Boettcher SW, Stucky GD, Moskovits M. Metal–silica hybrid nanostructures for surface-enhanced Raman spectroscopy. *Adv Mater.* 2006;18(21):2829–2832, doi: 10.1002/adma.200601254.
72. Cai H, Xu Y, Zhu N, He P, Fang Y. An electrochemical DNA hybridization detection assay based on a silver nanoparticle label. *Analyst.* 2002;127(6):803–808, doi: 10.1039/b200555g.
73. Luo X-L, Xu J-J, Zhao W, Chen H-Y. A novel glucose ENFET based on the special reactivity of MnO_2 nanoparticles. *Biosens Bioelectron.* 2004;19(10):1295–1300, doi: 10.1016/j.bios.2003.11.019.
74. Wang J, Liu G, Polsky R, Merkoçi A. Electrochemical stripping detection of DNA hybridization based on cadmium sulfide nanoparticle tags. *Electrochem Commun.* 2002;4(9):722–726, doi: 10.1016/S1388-2481(02)00434-4.
75. Huang Y-Y, et al. Luminescent supramolecular polymer nanoparticles for ratiometric hypoxia sensing, imaging and therapy. *Mater Chem Front.* 2018;2(10):1893–1899, doi: 10.1039/C8QM00309B.
76. Chakrabarty S, Maity S, Yazhini D, Ghosh A. Surface-directed disparity in self-assembled structures of small-peptide l-glutathione on gold and

silver nanoparticles. *Langmuir*. 2020;36(38):11255–11261, doi: 10.1021/acs.langmuir.0c01527.

77. Luo X-L, Xu J-J, Zhao W, Chen H-Y. A novel glucose ENFET based on the special reactivity of MnO_2 nanoparticles. *Biosensors and Bioelectronics*. 2004;19(10):1295–1300.
78. Luo X, Morrin A, Killard A-J, Smyth M-R. Application of nanoparticles in electrochemical sensors and biosensors. *Electroanalysis: An International Journal Devoted to Fundamental and Practical Aspects of Electroanalysis*. 2006;18(4):319–326.
79. Lin T-J, Huang K-T, Liu C-Y. Determination of organophosphorous pesticides by a novel biosensor based on localized surface plasmon resonance. *Biosens Bioelectron*. 2006;22(4):513–518, doi: 10.1016/j.bios.2006.05.007.
80. Okamoto T, Yamaguchi I, Kobayashi T. Local plasmon sensor with gold colloid monolayers deposited upon glass substrates. *Opt Lett*. 2000;25(6):372, doi: 10.1364/OL.25.000372.
81. Matsui J, et al. SPR sensor chip for detection of small molecules using molecularly imprinted polymer with embedded gold nanoparticles. *Anal Chem*. 2005;77(13):4282–4285, doi: 10.1021/ac050227i.
82. Qi Z, Honma I, Zhou H. Humidity sensor based on localized surface plasmon resonance of multilayer thin films of gold nanoparticles linked with myoglobin. *Opt Lett*. 2006;31(12):1854, doi: 10.1364/OL.31.001854.
83. Andreescu S, Luck LA. Studies of the binding and signaling of surface-immobilized periplasmic glucose receptors on gold nanoparticles: A glucose biosensor application. *Anal Biochem*. 2008;375(2):282–290, doi: 10.1016/j.ab.2007.12.035.
84. Turner M, et al. Selective oxidation with dioxygen by gold nanoparticle catalysts derived from 55-atom clusters. *Nature*. 2008;454(7207):981–983, doi: 10.1038/nature07194.
85. Dai ZH, Shao GJ, Hong JM, Bao JC, Shen J. Immobilization and direct electrochemistry of glucose oxidase on a tetragonal pyramid-shaped porous ZnO nanostructure for a glucose biosensor. *Biosens Bioelectron*. 2009;24(5):1286–1291.
86. Sousa CP, Polo AS, Torresi RM, De Torresi SIC, Alves WA. Chemical modification of a nanocrystalline TiO2 film for efficient electric connection of glucose oxidase. *J Colloid Interface Sci*. 2010;346(2):442–447.
87. Wang AJ, Li YF, Li ZH, Feng JJ, Sun YL, Chen JR. Amperometric glucose sensor based on enhanced catalytic reduction of oxygen using glucose oxidase adsorbed onto core-shell Fe3O4@silica@Au magnetic nanoparticles. *Mater Sci Eng C Mater Biol Appl*. 2012;32(6):1640–1647.

88. Zhao ZW, Chen XJ, Tay BK, Chen JS, Han ZJ, Khor KA. A novel amperometric biosensor based on ZnO: Co nanoclusters for biosensing glucose. *Biosens Bioelectron.* 2007;23(1):135–139.
89. Umar A, Rahman MM, Vaseem M, Hahn YB. Ultra-sensitive cholesterol biosensor based on low-temperature grown ZnO nanoparticles. *Electrochem Commun.* 2009;11(1):118–121.
90. Batra N, Tomar M, Gupta V. Realization of an efficient cholesterol biosensor using ZnO nanostructured thin film. *Analyst.* 2012;137(24):5854–5859.
91. Singh SP, Arya SK, Pandey P, Malhotra BD, Saha S, Sreenivas K, Gupta V. Cholesterol biosensor based on rf sputtered zinc oxide nanoporous thin film. *Appl Phys Lett.* 2007;91(6):063901.
92. Xiang C, Zou Y, Sun LX, Xu F. Direct electrochemistry and enhanced electrocatalysis of horseradish peroxidase based on flowerlike ZnO-gold nanoparticle-Nafion nanocomposite. *Sensors Actuators B.* 2009; 136(1):158–162.
93. Zhang Y, He PL, Hu NF. Horseradish peroxidase immobilized in TiO2 nanoparticle films on pyrolytic graphite electrodes: direct electrochemistry and bioelectrocatalysis. *Electrochim Acta.* 2004;49(12):1981–1988.
94. Zhang T, Tian BZ, Kong JL, Yang PY, Liu BH. A sensitive mediator-free tyrosinase biosensor based on an inorganic-organic hybrid titania sol-gel matrix. *Anal Chim Acta.* 2003;489(2):199–206.
95. Ansari SG, Wahab R, Ansari ZA, Kim YS, Khang G, Al HA, Shin HS. Effect of nanostructure on the urea sensing properties of sol-gel synthesized ZnO. *Sensors Actuators B.* 2009;137(2):566–573.
96. Zhao YG, Yan XQ, Kang Z, Lin P, Fang XF, Lei Y, Ma SW, Zhang Y. Highly sensitive uric acid biosensor based on individual zincoxide micro/nanowires. *Microchim Acta.* 2013;180:759–766.
97. Guan, H. N., Chi, D. F., Yu, J. Photoelectrochemical acetylcholinesterase biosensor incorporating zinc oxide nanoparticles. *Advanced Materials Research* (Vols. 183–185, pp. 1701–1706). 2011. Trans Tech Publications, Ltd. https://doi.org/10.4028/www.scientific.net/amr.183-185.1701
98. Mishra R, Dominguez R, Bhand S, Munoz R, Marty J. A novel automated flow-based biosensor for the determination of organophosphate pesticides in milk. *Biosens Bioelectron.* 2012;32:56–61.
99. Ghasemi-Varnamkhasti M, Rodriguez-Mendez ML, Mohtasebi SS, et al. Monitoring the aging of beers using a bioelectronic tongue. *Food Control.* 2012;25:216–224.
100. Monosik R, Stredansky M, Tkac J, Sturdik E. Application of enzyme biosensors in analysis of food and beverages enzyme and microbial technology. *Food Anal Methods.* 2012;5:40–53.

101. Arora P, Sindhu A, Dilbaghi N, Chaudhury A. Biosensors as innovative tools for the detection of food borne pathogens. *Biosens Bioelectron.* 2011;28:1–12.
102. Yan C, Dong F, Chun-yuan B, Si-rong Z, Jian-guo S. Recent progress of commercially available biosensors in china and their applications in fermentation processes. *J Northeast Agric Univ.* 2014;21:73–85.
103. Scognamiglio V, Pezzotti G, Pezzotti I, et al. Biosensors for effective environmental and agrifood protection and commercialization: from research to market. *Mikrochim Acta.* 2010;170:215–225.
104. Ooi KGJ, Galatowicz G, Towler HMA, Lightman SL, Calder VL. Multiplex cytokine detection versus ELISA for aqueous humor: IL-5, IL-10, and IFN profiles in uveitis. *Investig Ophthalmol Vis Sci.* 2006;47:272–277.
105. Caruso R, Trunfio S, Milazzo F, et al. Early expression of pro and anti-inflammatory cytokines in left ventricular assist device recipients with multiple organ failure syndrome. *Am Soc Art Int Org J.* 2010;56:313–318.
106. Wang H, Nakata E, Hamachi I. Recent progress in strategies for the creation of protein-based fluorescent biosensors. *Chem BioChem.* 2009;10:2560–2577.
107. Giuliano KA, Taylor DL. Fluorescent-protein biosensors: new tools for drug discovery. *Trends Biotechnol.* 1998;16:135–140.
108. Morris MC. Fluorescent biosensors – probing protein kinase function in cancer and drug discovery. *Biochim Biophys Acta.* 2013;1834:1387–1395.
109. Scognamiglio V, Arduini F, Palleschi G, Rea G. Biosensing technology for sustainable food safety. *Trends Anal Chem.* 2014;62:1–10.
110. German N, Ramanaviciene A, Voronovic J, Ramanavicius A. Glucose biosensor based on graphite electrodes modified with glucose oxidase and colloidal gold nanoparticles. *Mikrochim Acta.* 2010;168:221–229.
111. Backer D, Rakowski M, Poghossiana A, Biselli M, Wagner P, Schoning MJ. Chip-based amperometric enzyme sensor system for monitoring of bioprocesses by flow-injection analysis. *J Biotechnol.* 2013;163:371–376.
112. Amaro F, Turkewitz AP, Martin-Gonzalez A, Gutierrez JC. Whole-cell biosensors for detection of heavy metal ions in environmental samples based on metallothionein promoters from Tetrahymena thermophila. *Microb Biotechnol.* 2011;4:513–522.
113. Ivanov I, Younusov RR, Evtugyn GA, Arduini F, Moscone D, Palleschi G. Cholinesterase sensors based on screen-printed electrodes for detection of organophosphorus and carbamic pesticides. *Anal Bioanal Chem.* 2003;377:624–631.
114. Okumoto S. Quantitative imaging using genetically encoded sensors for small molecules in plants. *Plant J.* 2012;70:108–117.

115. Topell S, Glockshuber R. Circular permutation of the green fluorescent protein. *Methods Mol Biol.* 2002;183:31–48.
116. Bermejo C, Ewald JC, Lanquar V, Jones AM, Frommer WB. In vivo biochemistry: quantifying ion and metabolite levels in individual cells or cultures of yeast. *Biochem J.* 2011;438:1–10.
117. Woolston BM, Edgar S, Stephanopoulos G. Metabolic engineering: past and future. *Annu Rev Chem Biomol Eng.* 2013;4:259–288.
118. Becker K, Beer C, Freitag M, Kuck U. Genome-wide identification of target genes of a mating-type a-domain transcription factor reveals functions beyond sexual development. *Mol Microbiol.* 2015;96(5):1002–1022.
119. Berens C, Suess B. Riboswitch engineering – making the all important second and third steps. *Curr Opin Biotechnol.* 2015;31:10–15.

CHAPTER 6

Biosensor Overview and Its Advances for Cancer Detection

Prasanna Karki, Bibek Chettri, Pronita Chettri, Sanat Kr. Das, and Bikash Sharma

6.1 INTRODUCTION

Biosensors are advanced analytical tools that use biological sensing components to efficiently identify specific analytes. These devices consist of three core components: a bio-recognition element, a transducer, and an electronic system (Karunakaran et al. 2015). The bio-recognition element can include a range of biological components such as antibodies, DNA fragments, and enzymes, as well as nanomaterials like graphene, Transition Metal Dichalcogenides (TMDs), and MXenes (Naresh and Lee 2021; Arora 2013). The transducer turns biological responses into electrical signals that are then processed by an electronic system that includes a display, processor, and amplifier. A schematic illustration of a biosensor is shown in Figure 6.1. What sets biosensors apart is their ability to detect analytes and generate signals in proportion to their concentrations, operating on the principle of signal transduction. Because of their sensitivity, stability, affordability, and reproducibility, biosensors are extremely sought after.

Various sectors, including healthcare, agriculture, food manufacturing, and environmental monitoring, use biosensors (Dennison and Turner 1995; Baeumner et al. 2003; Geetha et al. 2006; Meshram et al. 2018; Chauhan et al. 2019; Mohankumar et al. 2021). In terms of selectivity and

DOI: 10.1201/9781003464211-6

FIGURE 6.1 Schematic illustration of a biosensor.

sensitivity, biosensors provide a level of sophistication that is superior to that of conventional diagnostic instruments. As technology has evolved, biosensors have transitioned from conventional to nanomaterial-based versions, leading to improvements in detection limits, sensitivity, and selectivity (Naresh and Lee 2021; Arora 2013). Their applications span a wide spectrum, from aiding in medical diagnosis to ensuring food safety and even addressing challenges in the marine industry. Biosensors play a crucial role in monitoring diseases, aiding in drug development, and detecting contaminants, microbes, and disease markers. Since their initiation by Leland C. Clark in 1962, biosensors have undergone remarkable scientific and practical advancements.

The classification of biosensors is based on the type of sensor device and biological materials they employ, with each category serving a specific purpose. Among these, electrochemical biosensors are among the most extensively studied and widely used. They operate by measuring the electrochemical characteristics of the analyte and the transducer, with subcategories including potentiometric, amperometric, impedimetric, and conductometric biosensors (Thévenot et al. 2001). Optical biosensors, on the other hand, integrate biological recognition elements into an optical transducer system, allowing real-time detection and offering both label-free and label-based types (Damborský et al. 2016). Electronic biosensors, based on field-effect transistors, use electric fields to control current flow, while thermal biosensors measure changes in thermal energy during biochemical reactions. Gravimetric biosensors are mass-based instruments detecting changes in mass due to binding interactions, and acoustic biosensors rely on alterations in acoustic wave characteristics due to absorbed analytes (Cali et al. 2020).

Moreover, biosensors can also be categorized based on the bioreceptors they employ. These categories include immunosensors, enzyme biosensors, DNA biosensors, and cell-based biosensors, each playing a unique role in diverse applications (Mosbach and Danielsson 1974; Roederer and

Bastiaans 1983; Wilson and Hu 2000; Banerjee and Bhunia 2010; Gui et al. 2017; Mishra et al. 2018; Asal et al. 2018; Singh et al. 2019; Campaña et al. 2019).

Biosensors are essential for environmental and health surveillance. They are excellent in environmental monitoring, contributing to the identification of contaminants and dangerous materials in the soil, water, and air, such as pesticides, poisons, carcinogens, mutagens, and endocrine-disrupting agents. They assist in food industry quality control systems and have been used in clinical practice, drug development, forensics, and biological research in the field of healthcare. Technological developments in biosensors have pushed the limits of sensitivity and greatly increased their efficiency. Various applications of biosensors are showcased in Figure 6.2.

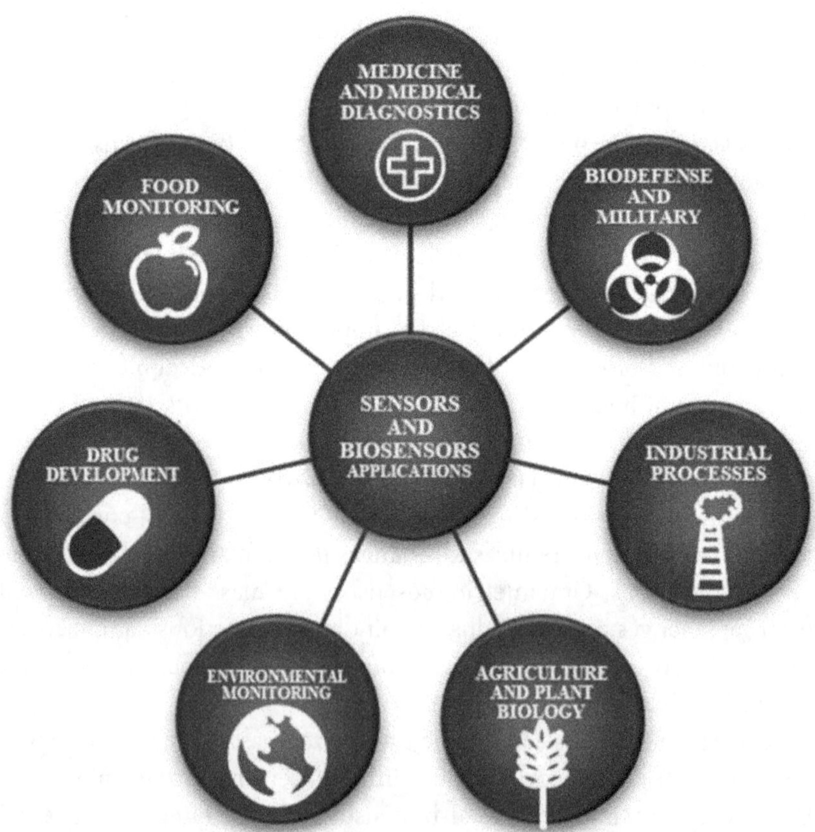

FIGURE 6.2 Application of biosensor (Singh et al. 2020).

Biosensors show tremendous potential in areas such as customized medicine, implantable devices for therapeutic monitoring, and real-time tracking of implanted medical equipment. Implantable biosensors have the potential to transform healthcare by continuously monitoring the impact of medications and therapies on the body, enabling tailored and precise healthcare approaches.

Nanotechnology is poised to play a major role in determining the future of biosensing technologies, promising additional breakthroughs in sensitivity and efficiency. As a result, biosensors are effective instruments that harness biological sensing elements and have the potential to improve the quality of life through a wide range of applications.

6.2 EVOLUTION OF BIOSENSOR TECHNOLOGY

The earliest known applications of biosensor technology date from the 1960s, when scientists started investigating the use of biological elements for sensing. A crucial moment occurred in 1962 with the creation of the oxygen electrode by Clark and Lyons, which could measure glucose levels in blood samples. This invention laid the foundation for successive developments in enzymatic biosensors. In the 1970s, Leland C. Clark Jr. introduced a revolutionary glucose oxidase-based biosensor that allowed people to monitor their blood sugar levels at home, transforming diabetes management and personalized healthcare (Bergveld 1970). Amperometric biosensors, which can detect a variety of analytes including cholesterol and neurotransmitters, were first introduced in the late 1980s, creating new opportunities for environmental monitoring and medical diagnostics (Cass et al. 1984). The early 1990s witnessed researchers exploring novel transducers and the potential of nanotechnology in biosensor development (Poncharal et al. 1999). Miniaturization, combined with improved sensitivity and selectivity through nanomaterials, made biosensors more user-friendly and portable.

The 2000s marked significant progress in biosensor technology, driven by advances in nanotechnology and materials science (Liu and Guo 2012; Tehrani et al. 2014; Walper et al. 2018; Basu et al. 2020). These developments enhanced the sophistication, sensitivity, and flexibility of biosensors. Among the notable accomplishments was the development of nanoscale biosensors that could monitor health in real time and identify poisons, heavy metals, and contaminants in the environment. Improvements in analytical performance became a priority, particularly in enhancing sensitivity and selectivity.

Nanomaterials like graphene and carbon nanotubes played a critical role in this pursuit due to their exceptional properties such as high surface area-to-volume ratio and excellent electrical conductivity (Basu and Bhattacharyya 2012; Justino et al. 2017). Additionally, the progress in data processing technology allowed for quicker analysis of the large datasets produced by biosensors. Machine learning algorithms were frequently employed to identify intricate patterns and recognize symptoms of disease. The development of portable and user-friendly point-of-care biosensors has received a lot of attention lately, allowing for quick diagnoses in a variety of situations without the need for sophisticated laboratory equipment. Personalized medicine has advanced greatly due to wearable biosensors that communicate with smartphones and other smart devices effortlessly. Looking to the future, biosensor technology will play an increasingly significant role, not only in healthcare but also in agriculture, food safety monitoring, and environmental conservation. The continuous progress of biosensors promises to enhance the quality of life and our understanding of the environment. The evolution of biosensors is shown in Figure 6.3.

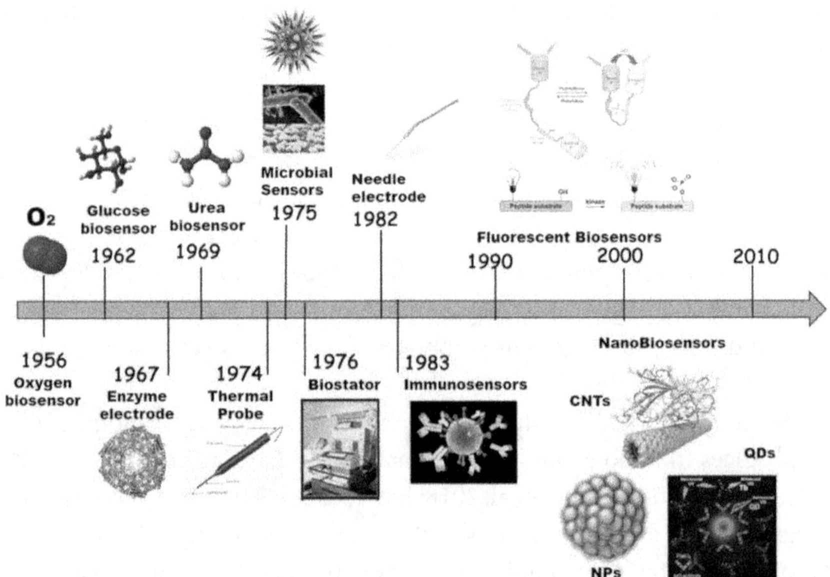

FIGURE 6.3 Evolution of biosensor (Tîlmaciu et al. 2015).

6.3 BIOSENSORS IN HUMAN HEALTH

Early disease detection is essential for efficient treatment and better patient outcomes. Conventional disease diagnosis techniques can entail invasive operations and drawn-out laboratory procedures. However, since the development of biosensor technology, the healthcare industry has undergone a significant change. Because biosensors can swiftly and accurately identify specific molecules or compounds in biological samples, they are revolutionizing the detection of disease biomarkers. The practice of finding molecules or substances that signal the existence of an illness, known as disease biomarker identification, has grown in significance within the healthcare industry. The potential of illness biomarker identification to support early diagnosis is a major factor in its importance. Healthcare practitioners can spot possible health problems before they develop into serious disorders by looking for biomarkers linked to diseases.

6.3.1 Nanomaterial in Biosensor

The development of sensor technology has rapidly evolved to meet the ever-expanding demands across various fields. Notably, as nanotechnology and nanoscience have advanced, sensor technology has gained significant popularity. The profound impact of nanotechnology has infused various scientific and industrial domains. The ability to manipulate and engineer materials at the atomic and molecular levels, coupled with the newfound insights into fundamental processes at the nanoscale, has opened exciting possibilities in the realm of biosensors. One critical factor in characterizing nanomaterials is their dimension. Nanomaterials are categorized based on their nanoscale dimensions: zero-dimensional (0D), one-dimensional (1D), two-dimensional (2D), and three-dimensional (3D) materials. A 0D nanomaterial refers to a substance that is nanoscale in all three dimensions. In contrast, 1D nanomaterial has two dimensions within the nanoscale range. 2D nanomaterials possess only one dimension within the nanoscale, and bulk nanomaterials extend beyond the nanoscale in all dimensions. In the context of our current epidemic, specifically, the need for efficient disease identification, research into enhancing the efficacy, sensitivity, and affordability of biosensors has prompted a keen interest in low-dimensional materials (LDM). These materials are currently being explored for their electrical and optical properties, with numerous innovative devices developed to leverage their unique atomic layer engineering capabilities. Researchers in the field of biosensing are increasingly turning their attention to nanomaterials like Fullerenes, magnetic nanoparticles (MNPs), graphene quantum dots (GQDs), carbon quantum dots (CQDs), inorganic quantum dots (QDs), and polymer dots. Among these, 0D nanomaterials have shown exceptional promise due to their minuscule size, quantum confinement effects, desirable physical and chemical properties, and excellent biocompatibility (Wang et al. 2020). For instance, GQDs have demonstrated tunable photoluminescence properties, making them valuable in bioimaging and biosensing (Sun et al. 2013). The presence of oxygen-rich functional groups at their edges enhances their water solubility and biocompatibility, while their anti-bleaching properties, luminescence stability, and high conductivity make GQDs-based sensors indispensable for diagnosing critical disorders and detecting a wide range of ions and biomarkers.

Carbon-based nanoparticles are highly researched due to their cost-effectiveness, low intrinsic toxicity, and adaptable surface functionalization, and have emerged as promising alternatives for various biological applications such as imaging, sensing, and drug delivery. Notably, 0D carbon nanomaterials exhibit superior electrical and optical properties, low toxicity, and high quantum yield, allowing for the construction of high-performance microsensors with minimal power consumption. Furthermore, polymer dots have gained traction as fluorescent probes for biosensors, particularly in biomolecular detection (Luo et al. 2019). Recent years have seen extensive research into 2D LDMs like graphene and biosensors based on TMDs for applications related to the detection of COVID-19 (Sharma et al. 2021; Afroj et al. 2021). While graphene-based biosensors have shown great promise, challenges related to bandgap and leakage current must be addressed to optimize their efficiency. LDMs like TMDs, MXenes, and Boron Nitride offer solutions to these issues and warrant further theoretical and practical exploration of current biosensors (Gupta et al. 2015).

Biosensors have been deployed for disease detection, health monitoring, heart disease diagnosis, optical diagnostics, and rapid DNA and RNA diagnostics (Hill et al. 2005). The application of biosensors in healthcare offers extensive benefits, from monitoring vital signs to tracking various diseases at home. Additionally, these sensors are valuable in environmental monitoring for detecting pollution's impact on the environment (Ahmed et al. 2019). In the food industry, biosensors expedite quality control processes, making them cost-effective and efficient. Further, biosensors are promising, with potential applications in precision medicine, implantable devices for therapeutic monitoring, and real-time tracking of implanted medical devices. These innovations will continue to revolutionize healthcare, environmental monitoring, food quality assurance, and more, cementing biosensors as an indispensable part of everyday life.

6.3.2 Biosensor for Cancer Detection

In the realm of cancer diagnosis, biosensors serve as essential tools for identifying specific tumor biomarkers. These biomarkers act as analytes, helping biosensors determine the presence or absence of a tumor and whether it is cancerous or non-cancerous (Parekh et al. 2018). Moreover, they play a crucial role in evaluating the effectiveness of cancer treatment by detecting protein levels expressed and/or secreted by tumor cells (Crulhas et al. 2017). Most cancers involve a multitude of biomarkers, and biosensors capable of detecting multiple analytes are exceptionally valuable for cancer monitoring and diagnosis (Hasanzadeh et al. 2017). Biosensors' ability to screen numerous markers not only aids in diagnosis but also saves time and resources (Bohunicky and Mousa 2011). These advanced biosensors often incorporate nanomaterials like carbon nanotubes, nanoparticles, and nanowires, which can be engineered to detect specific cancer biomarkers (Wang et al. 2017; Ivanov et al. 2018; Mahmoodi et al. 2020). Nanomaterials enhance biosensing sensitivity due to their unique optical and electrical properties, which can be functionalized to detect specific cancer biomarkers in samples. Nanowires, which are extremely thin wires with nanometer-scale diameters, are employed to detect very low concentrations of target molecules, enhancing biosensing specificity. The use of nanotechnologies has significantly advanced the field of biosensors for cancer detection. This progress allows for the development of highly sensitive and specific sensors that can detect cancer at early stages, potentially improving cancer diagnosis and treatment. Various biosensors utilizing nanomaterials and micro-technologies have been developed in the last decade for sensitive and specific cancer detection. For instance, in breast cancer detection, researchers have used nanowire biosensors based on silicon-on-insulator structures to detect microRNAs and DNA-oligonucleotides in the blood of breast cancer patients (Ivanov et al. 2018). They have also developed electrochemical molecularly imprinted polymer sensors for assessing HER2-ECD in breast cancer patients, providing accurate results (Pacheco et al. 2018; Soares et al. 2019). In prostate cancer, biosensors utilizing electrochemical impedance have been used to detect the PCA3 biomarker, demonstrating high sensitivity (Soares et al. 2019). A lectin-based capacitive non-faradaic biosensor has been employed for detecting the prostate-specific membrane antigen (PSMA), which is elevated in malignant lesions, aiding in distinguishing benign and malignant prostatic disorders (Subramani et al. 2021). Additionally, a nanostructured biosensor using two-dimensional layered nanomaterial $Ti_3C_2T_X$ has been created to detect sarcosine in urine, a potential biomarker for prostate cancer (Hroncekova et al. 2020). In cervical cancer detection, researchers have developed electrochemical DNA-based biosensors for the early detection of HPV-18 (Mahmoodi et al. 2020). Teengam et al. also reported paper-based graphene-polyaniline biosensors for human papillomavirus detection, offering a low-cost and disposable solution (Teengam et al. 2017). For ovarian cancer detection, biosensors using ZnO nanorods-Au nanoparticles nanohybrids have been fabricated to detect the ovarian cancer antigen (Gasparotto et al. 2017). In oral cancer detection, biosensors have been designed to detect biomarkers like CYFRA-21-1 and CD 59 in saliva samples (Choudhary et al. 2016; Kumar et al. 2019). Additionally, a fluorescence-based immunosensor utilizing carbon nanotubes has been developed for the salivary detection of oral squamous cell carcinoma. Furthermore, research is being directed to determine whether lung cancer can be detected through respired breath (Song et al. 2010; Buszewski et al. 2012; Saalberg and Wolff 2016; Liu et al. 2020; Karki et al. 2022). These biosensors offer quick and non-invasive methods for diagnosing cancer. These biosensor technologies hold promise for early cancer detection and diagnosis, potentially improving outcomes for cancer patients.

6.4 FUTURE PROSPECTS

In the future, biosensor techniques will investigate a growing range of opportunities for creating precise medications, tools, and diagnostics. The demand for creative approaches to the gathering of human samples is increasing. The development of tailored medicines could be greatly accelerated by implantable biosensors. By continuously monitoring the impacts of possible pharmaceuticals on the body, researchers may be able to fine-tune therapeutic techniques thanks to these sensors. Furthermore, the body may be implanted with biosensor chip technology to identify complicated blood DNA anomalies before the onset of disease symptoms. Real-time tracking of implanted medical devices and affordable, portable point-of-care devices are two potential uses for biosensor technology. Although biosensors have already had a significant impact in industries such as food, medicine, and environmental monitoring, their potential for addressing various challenges suggests that they will become an integral part of everyday life in the future.

6.5 CONCLUSION

Biosensors are clever gadgets that use live things to identify chemicals or traits in their surroundings. They are widely used in industrial, environmental, and clinical laboratories. Modern technology relies heavily on biosensors, which allow the identification of substances or traits without the need for separate cultivation, harvesting, or processing. These devices can function alone or as a component of automated testing systems. They are available in a wide variety. For many diseases to be successfully treated, early detection and diagnosis are essential, and biosensors present a viable answer. In the medical field, biosensors are widely used to help doctors and patients with diagnosis, clinical treatment, disease management, preventative measures, and health information. Nanomaterials have been increasingly prevalent in the development of biosensors in recent years. The several biosensor types and procedures based on analytes, transducers, and nanomaterials have been covered in this chapter. Because of the special qualities of nanomaterials, the integration of these materials has resulted in impressive advances in biosensing technology, improving biosensors' sensitivity, speed, response time, and repeatability.

REFERENCES

Afroj, Shaila, Liam Britnell, Tahmid Hasan, Daria V. Andreeva, Kostya S. Novoselov, and Nazmul Karim. 2021. "Graphene-Based Technologies for Tackling COVID-19 and Future Pandemics." *Advanced Functional Materials* 31 (52): 2107407. doi:10.1002/adfm.202107407.

Ahmed, Shadab, Naeem Shaikh, Nachiket Pathak, Akshay Sonawane, Vishal Pandey, and Samrat Maratkar. 2019. "An Overview of Sensitivity and Selectivity of Biosensors for Environmental Applications." *Tools, Techniques and Protocols for Monitoring Environmental Contaminants*, 53–73. Elsevier. doi:10.1016/B978-0-12-814679-8.00003-0.

Arora, N., 2013. "Recent Advances in Biosensors Technology: A Review." *Octa Journal of Biosciences*, 1(2): 147–150.

Asal, Melis, Özlem Özen, Mert Şahinler, and İlker Polatoğlu. 2018. "Recent Developments in Enzyme, DNA and Immuno-Based Biosensors." *Sensors* 18 (6): 1924. doi:10.3390/s18061924.

Baeumner, Antje J, Richard N Cohen, Vonya Miksic, and Junhong Min. 2003. "RNA Biosensor for the Rapid Detection of Viable Escherichia Coli in Drinking Water." *Biosensors and Bioelectronics* 18 (4): 405–413. doi:10.1016/S0956-5663(02)00162-8.

Banerjee, Pratik and Arun K. Bhunia. 2010. "Cell-Based Biosensor for Rapid Screening of Pathogens and Toxins." *Biosensors and Bioelectronics* 26 (1): 99–106. doi:10.1016/j.bios.2010.05.020.

Basu, Aviru Kumar, Adreeja Basu, and Shantanu Bhattacharya. 2020. "Micro/Nano Fabricated Cantilever Based Biosensor Platform: A Review and Recent Progress." *Enzyme and Microbial Technology* 139 (September): 109558. doi:10.1016/j.enzmictec.2020.109558.

Basu, S. and P. Bhattacharyya. 2012. "Recent Developments on Graphene and Graphene Oxide Based Solid State Gas Sensors." *Sensors and Actuators*, B: Chemical. doi:10.1016/j.snb.2012.07.092.

Bergveld, P. 1970. "Development of an Ion-Sensitive Solid-State Device for Neurophysiological Measurements." *IEEE Transactions on Biomedical Engineering BME* 17 (1): 70–71. doi:10.1109/TBME.1970.4502688.

Bohunicky, Brian and Shaker A. Mousa. 2011. "Biosensors: The New Wave in Cancer Diagnosis." *Nanotechnology, Science and Applications*. doi:10.2147/NSA.S13465.

Buszewski, Bogusaw, Agnieszka Ulanowska, Tomasz Kowalkowski, and Krzysztof Cieliski. 2012. "Investigation of Lung Cancer Biomarkers by Hyphenated Separation Techniques and Chemometrics." *Clinical Chemistry and Laboratory Medicine* 50 (3). doi:10.1515/cclm.2011.769.

Cali, Khasim, Elena Tuccori, and Krishna C. Persaud. 2020. "Gravimetric Biosensors." *Methods in Enzymology* 642: 435–468. doi:10.1016/bs.mie.2020.05.010.

Campaña, Ana, Sergio Florez, Mabel Noguera, Olga Fuentes, Paola Ruiz Puentes, Juan Cruz, and Johann Osma. 2019. "Enzyme-Based Electrochemical Biosensors for Microfluidic Platforms to Detect Pharmaceutical Residues in Wastewater." *Biosensors* 9 (1): 41. doi:10.3390/bios9010041.

Cass, Anthony E. G., Graham. Davis, Graeme D. Francis, H. Allen, O. Hill, William J. Aston, I. John. Higgins, Elliot V. Plotkin, Lesley D. L. Scott, and Anthony P. F. Turner. 1984. "Ferrocene-Mediated Enzyme Electrode for Amperometric Determination of Glucose." *Analytical Chemistry* 56 (4): 667–671. doi:10.1021/ac00268a018.

Chauhan, Nidhi, Utkarsh Jain, and Shringika Soni. 2019. "Sensors for Food Quality Monitoring." *Nanoscience for Sustainable Agriculture*, 601–626. Cham: Springer. doi:10.1007/978-3-319-97852-9_23.

Choudhary, Meenakshi, Prashant Yadav, Anu Singh, Satbir Kaur, Jaime Ramirez-Vick, Pranjal Chandra, Kavita Arora, and Surinder P. Singh. 2016. "CD 59 Targeted Ultrasensitive Electrochemical Immunosensor for Fast and Noninvasive Diagnosis of Oral Cancer." *Electroanalysis* 28 (10). doi:10.1002/elan.201600238.

Crulhas, Bruno P., Agnieszka E. Karpik, Flávia K. Delella, Gustavo R. Castro, and Valber A. Pedrosa. 2017. "Electrochemical Aptamer-Based Biosensor Developed to Monitor PSA and VEGF Released by Prostate Cancer Cells." *Analytical and Bioanalytical Chemistry* 409 (29). doi:10.1007/s00216-017-0630-1.

Damborský, Pavel, Juraj Švitel, and Jaroslav Katrlík. 2016. "Optical Biosensors." *Essays in Biochemistry* 60 (1): 91–100. doi:10.1042/EBC20150010.

Dennison, M.J., and A.P.F. Turner. 1995. "Biosensors for Environmental Monitoring." *Biotechnology Advances* 13 (1): 1–12. doi:10.1016/0734-9750(94)00020-D.

Gasparotto, Gisane, João Paulo C. Costa, Paulo I. Costa, Maria A. Zaghete, and Talita Mazon. 2017. "Electrochemical Immunosensor Based on ZnO Nanorods-Au Nanoparticles Nanohybrids for Ovarian Cancer Antigen CA-125 Detection." *Materials Science and Engineering C*. doi:10.1016/j.msec.2017.02.031.

Geetha, S., Chepuri R.K. Rao, M. Vijayan, and D.C. Trivedi. 2006. "Biosensing and Drug Delivery by Polypyrrole." *Analytica Chimica Acta* 568 (1–2): 119–125. doi:10.1016/j.aca.2005.10.011.

Gui, Qingyuan, Tom Lawson, Suyan Shan, Lu Yan, and Yong Liu. 2017. "The Application of Whole Cell-Based Biosensors for Use in Environmental Analysis and in Medical Diagnostics." *Sensors* 17 (7): 1623. doi:10.3390/s17071623.

Gupta, Ankur, Tamilselvan Sakthivel, and Sudipta Seal. 2015. "Recent Development in 2D Materials beyond Graphene." *Progress in Materials Science*. doi:10.1016/j.pmatsci.2015.02.002.

Halicka, Kinga, and Joanna Cabaj. 2021. "Electrospun Nanofibers for Sensing and Biosensing Applications—A Review." *International Journal of Molecular Sciences* 22 (12): 6357. doi:10.3390/ijms22126357.

Hasanzadeh, Mohammad, Nasrin Shadjou, and Miguel de la Guardia. 2017. "Non-Invasive Diagnosis of Oral Cancer: The Role of Electro-Analytical Methods and Nanomaterials." *TrAC - Trends in Analytical Chemistry*. doi:10.1016/j.trac.2017.04.007.

Hill, D., Chafin, D., Greco, R., Jafri, S., Murante, R., Noonan, J., Pham, A., Seabridge, S., Tannous, V., VanDerMeid, K., Wang, D., Wescott, N., McFarlane, K., Shah, S. 2005. "Development of a PCR free, fieldable, rapid, accurate, and sensitive bio-electronic DNA biosensor." *Proc. SPIE*, 5778, 283–292. https://doi.org/10.1117/12.606722

Hroncekova, Stefania, Tomas Bertok, Michal Hires, Eduard Jane, Lenka Lorencova, Alica Vikartovska, Aisha Tanvir, Peter Kasak, and Jan Tkac.

2020. "Ultrasensitive Ti3C2TX MXene/Chitosan Nanocomposite-Based Amperometric Biosensor for Detection of Potential Prostate Cancer Marker in Urine Samples." *Processes* 8 (5). MDPI AG. doi:10.3390/PR8050580.

Ivanov, Yu D., T. O. Pleshakova, K. A. Malsagova, A. F. Kozlov, A. L. Kaysheva, I. D. Shumov, R. A. Galiullin, et al. 2018. "Detection of Marker MiRNAs in Plasma Using SOI-NW Biosensor." *Sensors and Actuators, B: Chemical* 261. doi:10.1016/j.snb.2018.01.153.

Justino, Celine I.L., Ana R. Gomes, Ana C. Freitas, Armando C. Duarte, and Teresa A.P. Rocha-Santos. 2017. "Graphene Based Sensors and Biosensors." *TrAC - Trends in Analytical Chemistry*. doi:10.1016/j.trac.2017.04.003.

Karki, Prasanna, Bibek Chettri, Pronita Chettri, Sanat Kr. Das, and Bikash Sharma. 2022. "Computation Study of WSe2 Monolayer for Biomarker in Lung Cancer." *2022 IEEE International Conference of Electron Devices Society Kolkata Chapter (EDKCON)*, 371–374. IEEE. doi:10.1109/EDKCON56221.2022.10032891.

Karunakaran, Chandran, Raju Rajkumar, and Kalpana Bhargava. 2015. "Introduction to Biosensors." *Biosensors and Bioelectronics*, 1–68. Elsevier. doi:10.1016/B978-0-12-803100-1.00001-3.

Kumar, Suveen, Shweta Panwar, Saurabh Kumar, Shine Augustine, and Bansi D. Malhotra. 2019. "Biofunctionalized Nanostructured Yttria Modified Non-Invasive Impedometric Biosensor for Efficient Detection of Oral Cancer." *Nanomaterials* 9 (9). doi:10.3390/nano9091190.

Liu, Lian, Fang Lin, Xiaoting Ma, Zhaoxin Chen, and Jing Yu. 2020. "Tumor-Educated Platelet as Liquid Biopsy in Lung Cancer Patients." *Critical Reviews in Oncology/Hematology*. doi:10.1016/j.critrevonc.2020.102863.

Liu, Song, and Xuefeng Guo. 2012. "Carbon Nanomaterials Field-Effect-Transistor-Based Biosensors." *NPG Asia Materials* 4 (8): e23–e23. doi:10.1038/am.2012.42.

Luo, Jin-Hua, Qin Li, Shi-Hong Chen, and Ruo Yuan. 2019. "Coreactant-Free Dual Amplified Electrochemiluminescent Biosensor Based on Conjugated Polymer Dots for the Ultrasensitive Detection of MicroRNA." *ACS Applied Materials & Interfaces* 11 (30): 27363–27370. doi:10.1021/acsami.9b09339.

Mahmoodi, Pegah, Majid Rezayi, Elisa Rasouli, Amir Avan, Mehrdad Gholami, Majid Ghayour Mobarhan, Ehsan Karimi, and Yatima Alias. 2020. "Early-Stage Cervical Cancer Diagnosis Based on an Ultra-Sensitive Electrochemical DNA Nanobiosensor for HPV-18 Detection in Real Samples." *Journal of Nanobiotechnology* 18 (1). doi:10.1186/s12951-020-0577-9.

Meshram, B.D., A.K. Agrawal, Shaikh Adil, Suvartan Ranvir, and K.K. Sande. 2018. "Biosensor and Its Application in Food and Dairy Industry: A

Review." *International Journal of Current Microbiology and Applied Sciences* 7 (2): 3305–3324. doi:10.20546/ijcmas.2018.702.397.

Mishra, N, Y Kadam, and N Malek. 2018. "Overview of Enzyme Based Biosensors and Their Applications." *Current Trends in Biotechnology & Pharmacy* 12 (1): 108–117.

Mohankumar, P., J. Ajayan, T. Mohanraj, and R. Yasodharan. 2021. "Recent Developments in Biosensors for Healthcare and Biomedical Applications: A Review." *Measurement* 167 (January): 108293. doi:10.1016/j.measurement.2020.108293.

Mosbach, Klaus, and Bengt Danielsson. 1974. "An Enzyme Thermistor." *Biochimica et Biophysica Acta (BBA) - Enzymology* 364 (1): 140–145. doi:10.1016/0005-2744(74)90141-7.

Naresh, Varnakavi., and Nohyun Lee. 2021. "A Review on Biosensors and Recent Development of Nanostructured Materials-Enabled Biosensors." *Sensors* 21 (4): 1109. doi:10.3390/s21041109.

Pacheco, João G., Patrícia Rebelo, Maria Freitas, Henri P.A. Nouws, and Cristina Delerue-Matos. 2018. "Breast Cancer Biomarker (HER2-ECD) Detection Using a Molecularly Imprinted Electrochemical Sensor." *Sensors and Actuators, B: Chemical* 273. doi:10.1016/j.snb.2018.06.113.

Parekh, Aditya, Debanjan Das, Subhayan Das, Santanu Dhara, Karabi Biswas, Mahitosh Mandal, and Soumen Das. 2018. "Bioimpedimetric Analysis in Conjunction with Growth Dynamics to Differentiate Aggressiveness of Cancer Cells." *Scientific Reports* 8 (1). doi:10.1038/s41598-017-18965-9.

Poncharal, Philippe, Z. L. Wang, Daniel Ugarte, and Walt A. de Heer. 1999. "Electrostatic Deflections and Electromechanical Resonances of Carbon Nanotubes." *Science* 283 (5407): 1513–1516. doi:10.1126/science.283.5407.1513.

Roederer, Joy E., and Glenn J. Bastiaans. 1983. "Microgravimetric Immunoassay with Piezoelectric Crystals." *Analytical Chemistry* 55 (14): 2333–2336. doi:10.1021/ac00264a030.

Saalberg, Yannick, and Marcus Wolff. 2016. "VOC Breath Biomarkers in Lung Cancer." *Clinica Chimica Acta*. doi:10.1016/j.cca.2016.05.013.

Sharma, Sakshi, Sonakshi Saini, Maya Khangembam, and Vinod Singh. 2021. "Nanomaterials-Based Biosensors for COVID-19 Detection—A Review." *IEEE Sensors Journal* 21 (5): 5598–5611. doi:10.1109/JSEN.2020.3036748.

Singh, Ram Sarup, Taranjeet Singh, and Ashish Kumar Singh. 2019. "Enzymes as Diagnostic Tools." In *Advances in Enzyme Technology*, 225–271. Elsevier. doi:10.1016/B978-0-444-64114-4.00009-1.

Soares, Juliana Coatrini, Andrey Coatrini Soares, Valquiria Cruz Rodrigues, Matias Eliseo Melendez, Alexandre Cesar Santos, Eliney Ferreira Faria, Rui M. Reis, Andre Lopes Carvalho, and Osvaldo N. Oliveira. 2019. "Detection of the Prostate Cancer Biomarker PCA3 with Electrochemical and Impedance-Based Biosensors." *ACS Applied Materials and Interfaces* 11 (50). doi:10.1021/acsami.9b19180.

Song, Geng, Tao Qin, Hu Liu, Guo-Bing Xu, Yue-Yin Pan, Fu-Xing Xiong, Kang-Sheng Gu, Guo-Ping Sun, and Zhen-Dong Chen. 2010. "Quantitative Breath Analysis of Volatile Organic Compounds of Lung Cancer Patients." *Lung Cancer* 67 (2): 227–231. doi:10.1016/j.lungcan.2009.03.029.

Subramani, Indra Gandi, R. M. Ayub, Subash C.B. Gopinath, Veeradasan Perumal, M. F.M. Fathil, and M. K. Md Arshad. 2021. "Lectin Bioreceptor Approach in Capacitive Biosensor for Prostate-Specific Membrane Antigen Detection in Diagnosing Prostate Cancer." *Journal of the Taiwan Institute of Chemical Engineers* 120. doi:10.1016/j.jtice.2021.03.004.

Sun, Hanjun, Li Wu, Weili Wei, and Xiaogang Qu. 2013. "Recent Advances in Graphene Quantum Dots for Sensing." *Materials Today* 16 (11): 433–442. doi:10.1016/j.mattod.2013.10.020.

Teengam, Prinjaporn, Weena Siangproh, Adisorn Tuantranont, Charles S. Henry, Tirayut Vilaivan, and Orawon Chailapakul. 2017. "Electrochemical Paper-Based Peptide Nucleic Acid Biosensor for Detecting Human Papillomavirus." *Analytica Chimica Acta* 952. doi:10.1016/j.aca.2016.11.071.

Tehrani, Z., G. Burwell, M. A. Mohd Azmi, A. Castaing, R. Rickman, J. Almarashi, P. Dunstan, A. Miran Beigi, S. H. Doak, and O. J. Guy. 2014. "Generic Epitaxial Graphene Biosensors for Ultrasensitive Detection of Cancer Risk Biomarker." *2D Materials* 1 (2). IOP Publishing Ltd. doi:10.1088/2053-1583/1/2/025004.

Thévenot, Daniel R., Klara Toth, Richard A. Durst, and George S. Wilson. 2001. "Electrochemical Biosensors: Recommended Definitions and Classification1International Union of Pure and Applied Chemistry: Physical Chemistry Division, Commission I.7 (Biophysical Chemistry); Analytical Chemistry Division, Commission V.5 (Electroanalytical Chemistry).1." *Biosensors and Bioelectronics* 16 (1–2): 121–131. doi:10.1016/S0956-5663(01)00115-4.

Tîlmaciu, Carmen-Mihaela, and May C. Morris. 2015. "Carbon Nanotube Biosensors." *Frontiers in Chemistry* 3 (October). doi:10.3389/fchem.2015.00059.

Walper, Scott A., Guillermo Lasarte Aragonés, Kim E. Sapsford, Carl W. Brown, Clare E. Rowland, Joyce C. Breger, and Igor L. Medintz. 2018. "Detecting Biothreat Agents: From Current Diagnostics to Developing Sensor Technologies." *ACS Sensors* 3 (10): 1894–2024. doi:10.1021/acssensors.8b00420.

Wang, Baozhen, Uichi Akiba, and Jun Ichi Anzai. 2017. "Recent Progress in Nanomaterial-Based Electrochemical Biosensors for Cancer Biomarkers: A Review." *Molecules*. doi:10.3390/molecules22071048.

Wang, Zhengdi, Tingting Hu, Ruizheng Liang, and Min Wei. 2020. "Application of Zero-Dimensional Nanomaterials in Biosensing." *Frontiers in Chemistry* 8 (April). doi:10.3389/fchem.2020.00320.

Wilson, George S., and Yibai Hu. 2000. "Enzyme-Based Biosensors for in Vivo Measurements." *Chemical Reviews* 100 (7): 2693–2704. doi:10.1021/cr990003y.

CHAPTER 7

Recent Progress in Nanomaterial-Based IoT-Enabled Sensor

Rahul Samanta, Sandip Kunar, Gurudas Mandal, Atul Bandyopadhyay, Arindam Biswas, and Swarup Kumar Ghosh

7.1 INTRODUCTION

Demand for IoT-enabled chemical sensors with high performance has risen due to industrialization, where the need to monitor and prevent chemical hazardous species is crucial. Additionally, with the increased threat of viral infections, these sensors play a vital role in early detection and containment strategies [1]. Chemical sensors are gaining attention for their ability to detect biomarkers in the human body. This non-invasive approach proves invaluable for monitoring health conditions and diagnosing diseases [2–4]. Biomarkers can include gases, ions, and specific biocomponents such as glucose, viruses, and bacteria [5]. Chemical sensors are versatile and can detect various analytes, from gases and ions to biocomponents [6]. This versatility makes them applicable in both environmental monitoring and healthcare settings. Achieving high sensitivity and selectivity is crucial for detecting trace amounts of analytes [7]. These parameters ensure accurate and reliable sensor performance, enabling the detection of even minute quantities of target substances. The emphasis lies on swiftly detecting target analytes, especially to identify abnormal health states [8]. Rapid detection is essential for timely response and intervention. The low power consumption of sensor platforms is a need to enhance portability and enable on-site

DOI: 10.1201/9781003464211-7

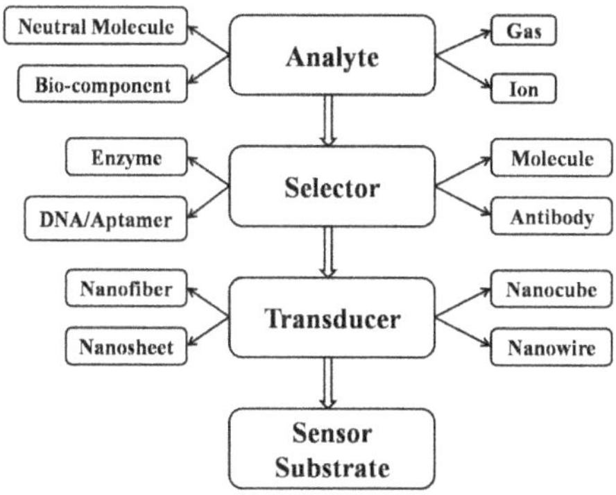

FIGURE 7.1 Schematic diagram of an IoT sensing platform.

detection [9]. These are particularly important representing the point-of-care testing (POCT), where tests are conducted, allowing immediate results [10, 11]. Figure 7.1 shows a schematic diagram [1] of an IoT sensing platform incorporated with nanomaterials.

The study highlighted the development of nanomaterials-based high-performance IoT-enabled chemical sensors driven by the need for efficient monitoring in industrial settings, the prevention of viral infections, and advancements in non-invasive healthcare diagnostics. The focus is on detecting various analytes with high sensitivity and selectivity, ensuring rapid and reliable results through miniaturization for portable applications. The primary emphasis is on nanomaterials, engineered at the structural level to enhance surface area and porosity, improving their sensitivity in chemical sensing applications. The primary objective for researchers is to develop next-generation chemical sensors by integrating novel IoT-enabled sensing materials into sensing systems [1, 12]. Moreover, structural engineering is employed to achieve large surface areas and high porosity, crucial for surface chemical reactions that form the basis of the sensing mechanism [13–15]. Nanomaterials undergo compositional and chemical reactivity modulations to tune their physical properties [16]. This tuning includes adjustments to electrical conductivity and optical emission, enhancing their capabilities as transducer layers on sensing substrates [17, 18]. Nanomaterials also serve as transducer layers on sensing substrates, producing distinguishable signal outputs [19]. Changes in the electrical and optical properties of the nanomaterials result in detectable signals, allowing for effective chemical sensing. Furthermore, to improve selectivity, specific binding is induced by functionalizing various selectors with nanomaterials [1, 20]. Selectors in this context may comprise synthetic molecules, antibodies, enzymes, and DNAs or aptamers [21, 22]. The study aims to present insight into developing next-generation chemical sensors using nanomaterials with engineered structures, enabling enhanced surface properties and sensitivity. Integrating these nanomaterials with IoT-enabled sensing systems, tuning their physical properties, and incorporating selective binding elements contribute to creating highly effective chemical sensing devices.

7.2 SYNTHESIS OF NANOMATERIALS AND ITS HYBRID STRUCTURE

7.2.1 Top-Down Approach

The top-down strategy is employed in the fabrication of nanomaterials, entailing the sequential manipulation to produce nanosized morphologies [23]. The process typically uses a ball mill, where the grinding medium (usually a liquid) and the material to be milled are placed in the mill, and the mechanical forces generated by the milling media lead to the reduction of the material into nanoscale particles [24]. This method is considered adequate for producing nanomaterials in large quantities due to its scalability and relatively straightforward process. It is essential in nanotechnology and materials science, where particle size control at the nanoscale is crucial for various applications, such as in electronics, medicine, and catalysis [25]. There are multiple techniques for producing nanoparticles and nanostructures, focusing on lithography, wet ball milling, and gas phase condensation. Some of the methods are discussed below:

a. ***Wet Ball Milling for Hexagonal Boron Nitride (h-BN) Nanoparticles:*** Wet ball milling has yielded non-crystalline or amorphous h-BN nanoparticles. Additionally, this method has been employed to produce nanostructures of boron nitride nanotubes (BNNTs), including bamboo-designed BN [26].
b. ***Condensation from the Gas Phase:*** In this approach, vaporized materials collide with inert gas molecules in the gaseous phase, leading to the dissipation of kinetic energy and the subsequent condensation into nanoscopic crystals. The condensed crystals accumulate on a substrate as extremely fine powder [27].

c. ***Copper Nanoparticles Condensation and Molecular Dynamic Modeling:*** Chepkasov et al. [28] conducted an experimental investigation on the condensation of copper nanoparticles from the gas phase. Molecular dynamic modeling was employed, utilizing a digital model comprising 8500 copper atoms to accurately depict vaporization and condensation mechanisms.
d. ***Lithography for 1D Nanostructures:*** Lithography is a versatile and straightforward approach for producing self-assembled one-dimensional (1D) nanostructures on diverse substrates [29]. This method is rapid and efficient for surface mapping and is applicable across a broad spectrum of substrates.
e. ***Development of Lithographic Techniques:*** Different lithographic techniques, such as extreme ultraviolet (EUV) and X-ray lithography (XRL), are currently under development [30, 31]. XRL has been extensively investigated utilizing radiations within the wavelength range of 0.1–10 nm, particularly for high-resolution applications.

7.2.2 Bottom-Up Approach

This is favored for the preparation of nanostructures because it allows for the creation of materials with unique interdisciplinary characteristics that are different from those demonstrated by the individual building constituents [32]. The methods mentioned, particularly CVD and MBE, exemplify the bottom-up approach by utilizing processes involving controlled interactions and atomic or molecular depositions [33]. These techniques offer advantages in terms of precision and control, making them valuable in various applications, especially in the semiconductor industry [34].

a. ***CVD Method:*** CVD is prominently recognized as a widely employed method in the semiconductor industry for depositing thin layers on various substrates [35]. These precursors, when decomposed, interact with each other to produce the desired deposition. The volatilized precursors initially adsorb on a surface at high temperatures, leading to interactions or disintegration and the formation of crystals.
b. ***MBE Method:*** MBE is presented as another method of physical evaporation that differs from conventional epitaxy systems. Unlike methods relying on chemical interactions, MBE utilizes a simple

physical evaporation process. In MBE, material is deposited atom by atom or molecule by molecule in a controlled environment, allowing precise control over the growth of thin films. Notably, this process effectively operates at lower temperatures compared to vapour-phase epitaxy. It provides practical benefits, including precise control over the growth of thin films and the ability to produce well-defined nanostructures [36].

c. ***MSA Method:*** It is a noncovalent binding process where atoms and molecules assemble into a stable and distinct nanophase, excelling in producing nanoparticles in the 1–100 nm range [37].

d. ***Sol-Gel Method:*** The method involves merging dispersed solid nanoparticles (sols) with diameters ranging from 1 to 100 nm in a homogeneous liquid medium [38].

e. ***1D TiO_2 Nanostructures:*** These methods have been instrumental in producing various nanostructures, including nanorods, nanowires (NWs), nanotubes, and nanobelts [39]. These nanostructures are utilized in optoelectronic and gas sensor applications.

7.2.2.1 Nanotubes

Carbon nanotubes (CNTs) have unique structural qualities and properties that have sparked interest due to their potential technological applications. CNTs have a cylindrical structure constructed from a hexagonal carbon lattice, resembling a rolled-up graphite sheet [40]. Nanotubes showcase diverse electrical, structural, and thermal properties, contingent on factors such as the nanotube's length, diameter, and chirality (twist) [41]. They can exist in various forms, including single-walled nanotubes (SWNTs) with a single cylindrical wall and multi-walled nanotubes (MWNTs) with multiple cylinders nested within one another. The synthesis of CNTs can employ various methods and precursor materials. The researchers [42] illustrated the pyrolysis of suitable components to generate boron carbide nitride (B–C–N) and C–N nanotubes. Pyrolyzing aza-aromatics, such as pyridine, over cobalt catalysts, can result in the production of CN nanotubes [43, 44]. Overall, the diverse properties and potential applications of CNTs make it a fascinating area of study in materials science and nanotechnology. The synthesis methods control the properties of CNTs for various applications in electronics, materials engineering, and other fields.

7.2.2.2 Nanowires

Semiconducting NWs can be fabricated using various physicochemical methods. The strategies for manufacturing semiconducting NWs can be categorized into bottom-up and top-down approaches [45]. Using high-energy plasma species, NWs are created through a physical process that involves cutting bulk single-crystalline materials (e.g., Si, Ge, GaAs). The VLS growth process uses catalyst particles, initially reported with Au-catalyzed Si micro-whiskers in 1964 [46]. Furthermore, conventional approaches for establishing the epitaxial structure via the VLS method enable precise alignment during NW formation [47]. Similar epitaxial control levels are attainable with materials such as GaN and Si/Ge structures [48]. The VLS approach facilitates the creation of NW structures with precise shape control (<20 nm) [49]. The TEM has played a crucial role in characterizing these materials, paving the way for advancements in advanced materials research [50, 51].

7.2.2.3 Nanorods

The development of nanorods has garnered significant interest in various research fields and applications. The unique properties of nanorods, stemming from their form anisotropy, make them appealing alternatives for multiple uses. One notable advantage is their increased efficiency compared to spherical particles, primarily attributed to the heightened stimulation of surface plasmons resulting from an increase in the aspect ratio of the particles. The dipole moment strength within a nanoparticle becomes notably significant due to the rise in plasmons on the surface. This results in an enhancement of the electrical field in nanorods compared to spherical particles. Well-oriented CdSe nanorods have been proven effective as a guided path for charge transporters in photoelectric devices, thereby contributing to improved collection efficiency. Moreover, the incorporation of nanorods into P3HT films has substantially increased the extrinsic quantum yield, with a threefold improvement observed when the aspect ratio was elevated from 1 to 10 nm [52]. Notably, the benefits observed in the inherent properties of nanorods are expected to extend to larger rod-shaped particles, encompassing variations in both diameter and length [53]. Moreover, the efficacy of nanorods is significantly influenced by factors such as aspect ratio, polydispersity, volume fraction, and alignment. This understanding contributes to their potential applications across diverse fields, including electronics, optics, and materials science.

7.3 SEVERAL TYPES OF IOT-ENABLED SENSORS

7.3.1 Biosensors

In the modern day, advancements in biosensor development, particularly in the context of POCT, are used for the rapid and precise detection of biological signals. Several innovations have been made in integrating biological elements with signal transducers, creating a new generation of sensor systems. Innovations in biosensor development involve combining biological elements with signal transducers [54]. The sensor systems are becoming more advanced, flexible, and wearable. Integration with IoT sensing platforms provides for enhanced functionality and connectivity. These sensor systems are specifically engineered to detect various biomolecules in the human body, including glucose, lactate, uric acid, and bacteria (e.g., pathogenic *Escherichia coli*). Novel glucose sensors are garnering substantial attention worldwide [55]. Early diagnosis of diabetes is emphasized, considering the projected increase in the global diabetes population. Chronic hyperglycaemia in diabetes can lead to uncontrolled blood glucose levels, contributing to various complications, including blindness, nerve damage, cardiovascular disease, and kidney failure [56]. Proactive measures, including continuous real-time monitoring of blood glucose levels, are essential for providing adequate medical treatment and preventing diabetes-related complications [57]. In addition, developing advanced biosensors, especially those integrated with flexible and wearable systems, contributes to improving POCT for the rapid and precise detection of various biomolecules, focusing on glucose sensors for non-invasive diabetes diagnosis. Continuous real-time monitoring is crucial for early diagnosis and effective diabetes management.

7.3.2 Gas Sensors

Researchers are significantly focusing on integrating nanomaterials with IoT-based gas-sensing modules for diverse applications, including monitoring hazardous environments, assessing food freshness, and aiding in disease diagnosis [58]. The detection of nitrogen dioxide (NO_2) serves as an illustrative example, underscoring the importance of monitoring toxic gases emitted from sources such as automobiles and industrial plants, which can contribute to respiratory diseases [59]. The potential of highly sensitive gas sensors in POCT applications become a key prospect nowadays. Moreover, the nanostructure, such as nanorods, NWs, nanofibers

(NFs), and nanotubes, as gas-sensing layers, is essential to serve the purpose [60]. These structures offer advantages like large surface area and porosity, contributing to high sensitivity in gas detection [61].

7.3.3 Ion Sensors

The primary objective is to enable real-time analysis through the swift detection of analyte species, minimize sensing platforms, and achieve quantitative analysis of ion concentrations [62]. Wearable sensing platforms, particularly for analysing biofluids like sweat, offer non-invasive healthcare monitoring and POCT [63]. Acetate is highlighted as a metabolic switch controlling bacterial cell growth, with implications for applications such as the production of recombinant proteins. An increased chloride concentration in sweat is identified as a biomarker for diagnosing cystic fibrosis. These receptors are designed with unique geometries to form stable six- and eight-membered chelated structures with halides and Y-shaped oxoanions [64]. The acidity of the receptors plays a role in determining the specific binding mechanism [65]. The advancements in ion sensor systems, with a primary focus on anion detection and the design of receptors with dual-hydrogen bond donors for applications in healthcare and environmental monitoring, have become a significant demand to enhance the early detection of diseases.

7.4 APPLICATIONS OF NANOMATERIALS FOR IOT-ENABLED SENSORS

7.4.1 Gas Sensing

7.4.1.1 Automobile Exerted Gases Safety Monitoring

Recent research has revealed the importance of gas sensors in vehicles, mainly focusing on their role in monitoring combustible gases and contributing to air quality control. It has been highlighted that the increasing integration of gas sensors into modern vehicles' electronic control units to regulate air intake and maintain air quality standards [66]. The significance of monitoring gases such as CO and nitrogen dioxide NO_2 gas, which are indicators of combustion, becomes necessary. Integrating these sensitive sensors becomes a common practice in automobiles to ensure compliance with air quality regulations. Furthermore, the automotive industry aims to decrease atmospheric contamination and improve fuel efficiency [67]. Gas sensors play a crucial role in achieving safety goals by offering design flexibility and high sensitivity. The introduction of resistive gas sensors based on nanomaterials is seen as a solution for monitoring common gases [68]. An illustrative example involves the use of Nb NFs as a material for constructing a resistive gas sensor designed for room-temperature CO detection. This example highlights the sensor's impressive response percentages for different concentrations of CO.

7.4.1.2 Monitoring of Health

Various gas sensors, especially those based on semiconductor technology, play a crucial role in maintaining safety in various industries by detecting and alerting personnel to the presence of harmful gases. This is particularly important in environments where dangerous gases are commonly released, such as underground mining, gas and oil industries, and chemical facilities [69]. Efficient gas-sensing systems contribute significantly to the safety and well-being of individuals in these sectors and those residing nearby. Underground miners face various risks, such as combustible gases, asphyxiates, and low oxygen levels. Gas sensors, typically semiconductors, are employed to detect these gases. These gases may be encountered during production, transportation, processing, storage, or in proximity to distribution pipes [70]. Gas sensors help mitigate the risks associated with the movement of dangerous gases. Chemical facilities across different sectors release hazardous gases as by-products during their processes. The release of these gases poses a threat to individuals living in proximity to such facilities [71]. Highly efficient gas-sensing systems are imperative in these sectors to monitor and manage gas emissions effectively.

7.4.1.3 Monitoring of Environmental Hazards

Three categories of gaseous contaminants are identified: natural, synthetic, and anthropogenic gases [72]. Natural causes of air pollution, including volcano eruptions, wildfires, and lightning, are contrasted with synthetic sources, which arise from human activities such as automobile exhaust gases, chemical disasters, and commercial activities. The impact of anthropogenic activities on climate change increases the need for monitoring pollutants [73]. Using gas sensors to screen gases like CO_2, NO_2, SO_2, and O_3 can be beneficial to achieve this. The researchers developed an edge-exposed WS_2 (tungsten disulphide) on SiO_2 NRs (silicon dioxide nanorods) capable of accurately and precisely detecting NO_2. The response of WS_2 was particularly significant at 5 ppm NO_2, surpassing reactions to interfering gases at ambient temperature [74].

7.4.2 Optoelectronic Application

7.4.2.1 Photodetector

The potential of hybrid nanostructures in improving the efficiency and economics of nano photodetectors and photocatalysts, with specific enhanced performance through modifications and additions of nanomaterials, has become a major prospect [75]. The bandgap of semiconductors varies, with CdTe having a bandgap of 1.5 eV and ZnS having a bandgap of 3.7 eV. This property is crucial for their application in nano photodetectors. Nanostructures are considered promising for applications such as nano photodetectors or nano-switches that operate in the near-infrared (NIR) to ultraviolet (UV) range. Nanostructures are believed to offer greater efficiency and economic benefits than traditional film and bulk devices due to their unique properties, including higher absorption coefficients that enhance their performance [76]. Composite NW photocatalysts have been developed as a proof-of-concept, demonstrating increased optical emission in modified nanomaterials. For example, CdSe nanoribbons decorated with Au hollow nanoparticles improve photodetection effectiveness compared to bare CdSe nanoribbons. Various nanostructures, such as Au-decorated CdSe and CdTe NWs, Au-decorated ZnO/ZnCdSeTe composite nanowires (CSNWs), Ag-decorated ZnO nanorod array light-emitting diodes (LEDs), and Si NW arrays covered with Au nanomaterial-coated graphene sheets, have been demonstrated as superior in plasmon-enhanced light transfer efficiency [77]. This suggests that incorporating noble metal nanoparticles (e.g., Au, Ag) into nanostructures can improve their performance in light absorption and emission.

7.4.2.2 Solar Cell

Bulk semiconductors in the II–VI groups, particularly CdTe, are utilized as viable candidates for solar applications. Additionally, the potential of heterojunctions in improving solar cells by integrating materials with distinct bandwidths to absorb sunlight in specific spectral regions preferentially also grabs the attention of researchers [78]. Perovskite-based materials, such as CsPbI3, are also mentioned for their widespread use in photovoltaics (PVs) despite concerns about their stability. The optimization of light absorption is highlighted through strategies like increasing the interfacial region and illumination entrapment. TiO_2–SiO_2 comprises the cubic perovskite phase in the atmosphere and finds extensive applications [79]. Furthermore, solar systems were developed using CdS, Ga nanoribbons, and Si heterojunctions, demonstrating open circuit photovoltage, short-circuit current, throughput, and energy transfer effectiveness.

7.4.2.3 Light-Emitting Diode

In this study, the potential of hybrid perovskites in achieving high photovoltaic performance and their applicability in light-emitting devices have been described. The advancements in building electroluminescent diodes, such as micro-LEDs, using semiconductor nanostructures are gaining attention, and they can be explained by the Shockley-Queisser theory [80]. The Shockley-Queisser limit is a theoretical efficiency limit for a solar cell, representing the maximum attainable efficiency that a solar cell can achieve under ideal conditions. The bandgap of the material influences the limit, and it provides insight into the maximum efficiency achievable with a particular semiconductor material. In contrast, hybrid perovskites refer to materials used in photovoltaic applications. They typically consist of organic and inorganic components arranged in a perovskite crystal structure [81]. Hybrid perovskite solar cells have gained attention for their excellent optoelectronic properties, making them promising candidates for efficient and low-cost solar energy conversion. Moreover, photoluminescence (PL) and quantum yield refer to the efficiency of light emission in a material. In hybrid perovskites, a high photoluminescence quantum yield indicates that a significant portion of absorbed light is re-emitted as luminescence rather than lost through non-radiative processes [82]. High quantum yields are desirable for efficient light absorption and emission in solar cells and light-emitting devices. Micro-LEDs are miniature LEDs that have dimensions in the micrometre range. They are used in displays, lighting, and other optoelectronic devices. The development of electronically driven electroluminescent diodes, including micro-LEDs, using various semiconductor nanostructures can effectively lead the world to a better place.

7.5 FUTURE PERSPECTIVES/CHALLENGES AND THEIR MITIGATION PROCESSES

a. Stability of Nanomaterials

   Challenge: Nanomaterials can be susceptible to degradation or instability over time, affecting the sensor's performance.

   Solution: Further research and development are needed to enhance nanomaterials' stability, possibly through protective coatings or improved synthesis methods.

b. Rapid On-Site Detection of Analytes

Challenge: Achieving fast and accurate on-site detection is crucial for many applications, and current technologies may have limitations in terms of speed.

Solution: Developing nanomaterial-based sensors with faster response times and efficient detection mechanisms can address this challenge.

c. *Sensor Reliability*

Challenge: Ensuring the reliability and reproducibility of sensor readings is essential for practical applications.

Solution: Rigorous testing, quality control measures, and standardization of nanomaterial-based sensors can improve their reliability.

d. Low Power Consumption

Challenge: Many applications require sensors with low power consumption to enable long-term and portable use.

Solution: Designing energy-efficient sensing platforms and exploring low-power electronics can contribute to addressing this challenge.

e. *Portability*

Challenge: Some sensing platforms may lack portability, hindering their widespread use in various settings.

Solution: Development of compact and portable sensing devices, possibly leveraging advancements in miniaturization and integration technologies.

f. *Usability of Sensing Platforms*

Challenge: Ensuring that sensing platforms are user-friendly is crucial for widespread adoption.

Solution: Human-centred design principles can be applied to create intuitive interfaces and simplify the operation of nanomaterial-based sensors.

g. *Integration with Mobile Platforms for Continuous Monitoring*

Challenge: Seamless integration of nanomaterial-based sensors with mobile devices for continuous monitoring poses technical and logistical challenges.

Solution: Continued efforts to develop user-friendly mobile applications and communication protocols that enable real-time data transfer from sensors to mobile platforms.

h. *Chemical Sensors for POCT Based on IoT*

Challenge: Meeting the increasing demand for effective IoT-based chemical sensors for POCT.

Solution: Ongoing research to develop innovative IoT-based solutions that offer rapid and reliable chemical sensing for personal healthcare and environmental monitoring.

Addressing these challenges requires interdisciplinary efforts involving materials science, engineering, and data science. Continued research and innovation will contribute to developing advanced nanomaterial-based sensing platforms with improved properties and broader applications.

7.6 CONCLUDING REMARKS

The key focus of this study is on the current advancements in the fabrication, characterization, and implementation of nanomaterials incorporated with IoT sensors. These structures can be created using either bottom-up or top-down techniques, and they effectively hold great promise for applications like gas censoring, optoelectronic devices, and many more. The unique properties of the nanomaterials, such as their narrow shape and large surface area, make them particularly suitable for these purposes. Despite significant research efforts in nanostructures, some challenges must be addressed before these materials can be effectively commercialized. One of the key challenges highlighted is the scaling up of manufacturing processes. While these nanostructures may exhibit favourable characteristics at the laboratory scale, achieving consistent quality and maintaining those characteristics during mass production is a complex task. For example, in the case of lithography, EUV lithography faces challenges due to restricted power in EUV sources. The difficulties in scaling up production may arise from various factors, including the need for precise control over the fabrication process, ensuring reproducibility, and managing the potential variations at larger production scales. Also, maintaining the nanostructure's desired properties throughout mass production is crucial for their successful commercialization. Moreover, the chapter acknowledges the current nanostructure achievements for applications in various IoT-enabled censoring systems.

REFERENCES

[1] Choi, S.H., Lee, J.S., Choi, W.J., Seo, J.W. and Choi, S.J., 2022. Nanomaterials for IoT sensing platforms and point-of-care applications in South Korea. *Sensors*, 22(2), p. 610.

[2] Lim, J.W., Kim, T.Y. and Woo, M.A., 2021. Trends in sensor development toward next-generation point-of-care testing for mercury. *Biosensors and Bioelectronics*, 183, p. 113228.

[3] Yoon, J.W. and Lee, J.H., 2017. Toward breath analysis on a chip for disease diagnosis using semiconductor-based chemiresistors: Recent progress and future perspectives. *Lab on a Chip*, 17(21), pp. 3537–3557.

[4] Sempionatto, J.R., Jeerapan, I., Krishnan, S. and Wang, J., 2019. Wearable chemical sensors: emerging systems for on-body analytical chemistry. *Analytical chemistry*, 92(1), pp. 378–396.

[5] Justino, C.I., Freitas, A.C., Pereira, R., Duarte, A.C. and Santos, T.A.R., 2015. Recent developments in recognition elements for chemical sensors and biosensors. *TrAC Trends in Analytical Chemistry*, 68, pp. 2–17.

[6] Majhi, S.M., Mirzaei, A., Kim, H.W., Kim, S.S. and Kim, T.W., 2021. Recent advances in energy-saving chemiresistive gas sensors: A review. *Nano Energy*, 79, p. 105369.

[7] Choi, S.J. and Kim, I.D., 2018. Recent developments in 2D nanomaterials for chemiresistive-type gas sensors. *Electronic Materials Letters*, 14, pp. 221–260.

[8] Samanta, R., Bandyopadhyay, A., Mondal, A., Das, A., Sinha, A. and Mandal, G., 2023. Recent Trends in Nanometric Dispersed Polymer Composites. In *Polymer Nanocomposites* (pp. 109–120). CRC Press.

[9] Yang, C., Denno, M.E., Pyakurel, P. and Venton, B.J., 2015. Recent trends in carbon nanomaterial-based electrochemical sensors for biomolecules: A review. *Analytica Chimica Acta*, 887, pp. 17–37.

[10] Yao, S., Swetha, P. and Zhu, Y., 2018. Nanomaterial-enabled wearable sensors for healthcare. *Advanced Healthcare Materials*, 7(1), p. 1700889.

[11] Kim, S.J., Choi, S.J., Jang, J.S., Cho, H.J. and Kim, I.D., 2017. Innovative nanosensor for disease diagnosis. *Accounts of Chemical Research*, 50(7), pp. 1587–1596.

[12] Wang, Z., Hu, T., Liang, R. and Wei, M., 2020. Application of zero-dimensional nanomaterials in biosensing. *Frontiers in Chemistry*, 8, p. 320.

[13] Choi, S.J., Persano, L., Camposeo, A., Jang, J.S., Koo, W.T., Kim, S.J., Cho, H.J., Kim, I.D. and Pisignano, D., 2017. Electrospun nanostructures for high performance chemiresistive and optical sensors. *Macromolecular Materials and Engineering*, 302(8), p. 1600569.

[14] Choi, S.J., Kim, I.D. and Park, H.J., 2022. 2D layered Mn and Ru oxide nanosheets for real-time breath humidity monitoring. *Applied Surface Science*, 573, p. 151481.

[15] Liu, X., Ma, T., Pinna, N. and Zhang, J., 2017. Two-dimensional nanostructured materials for gas sensing. *Advanced Functional Materials*, 27(37), p. 1702168.

[16] Jeong, J.M., Yang, M., Kim, D.S., Lee, T.J. and Choi, B.G., 2017. High performance electrochemical glucose sensor based on three-dimensional MoS2/graphene aerogel. *Journal of Colloid and Interface Science*, 506, pp. 379–385.

[17] Choi, S.J., Choi, H.J., Koo, W.T., Huh, D., Lee, H. and Kim, I.D., 2017. Metal–organic framework-templated PdO-Co3O4 nanocubes functionalized by SWCNTs: Improved NO_2 reaction kinetics on flexible heating film. *ACS Applied Materials & Interfaces*, 9(46), pp. 40593–40603.

[18] Kwon, O.S., Song, H.S., Park, T.H. and Jang, J., 2018. Conducting nanomaterial sensor using natural receptors. *Chemical Reviews*, 119(1), pp. 36–93.

[19] Fang, Y., Deng, Y. and Dehaen, W., 2020. Tailoring pillararene-based receptors for specific metal ion binding: From recognition to supramolecular assembly. *Coordination Chemistry Reviews*, 415, p. 213313.

[20] Durkin, T.J., Barua, B. and Savagatrup, S., 2021. Rapid detection of sepsis: Recent advances in biomarker sensing platforms. *ACS Omega*, 6(47), pp. 31390–31395.

[21] Kucherenko, I.S., Soldatkin, O.O., Dzyadevych, S.V. and Soldatkin, A.P., 2020. Electrochemical biosensors based on multienzyme systems: main groups, advantages and limitations–a review. *Analytica Chimica Acta*, 1111, pp. 114–131.

[22] Li, F., Yu, Z., Han, X. and Lai, R.Y., 2019. Electrochemical aptamer-based sensors for food and water analysis: A review. *Analytica Chimica Acta*, 1051, pp. 1–23.

[23] Rawat, J., Sharma, H. and Dwivedi, C., 2023. One-dimensional semiconducting hybrid nanostructure: Gas sensing and optoelectronic applications. *1D Semiconducting Hybrid Nanostructures: Synthesis and Applications in Gas Sensing and Optoelectronics*. WILEY-VCH GmbH. pp. 1–26.

[24] Sotiropoulou, S., Sierra-Sastre, Y., Mark, S.S. and Batt, C.A., 2008. Biotemplated nanostructured materials. *Chemistry of Materials*, 20(3), pp. 821–834.

[25] Arole, V.M. and Munde, S.V., 2014. Fabrication of nanomaterials by top-down and bottom-up approaches-an overview. *J. Mater. Sci*, 1, pp. 89–93.

[26] Chen, Y., Chadderton, L.T., Gerald, J.F. and Williams, J.S., 1999. A solid-state process for the formation of boron nitride nanotubes. *Applied Physics Letters*, 74(20), pp. 2960–2962.

[27] Fernández, A., Reddy, E.P., Rojas, T.C. and Sánchez-López, J.C., 1999. Application of the gas phase condensation to the preparation of nanoparticles. *Vacuum*, 52(1–2), pp. 83–88.

[28] Chepkasov, I.V., Gafner, Y.Y., Gafner, S.L. and Bardahanov, S.P., 2016. Condensation of Cu nanoparticles from the gas phase. *The Physics of Metals and Metallography*, 117, pp. 1003–1012.

[29] Colson, P., Henrist, C. and Cloots, R., 2013. Nanosphere lithography: a powerful method for the controlled manufacturing of nanomaterials. *Journal of Nanomaterials*, 2013, 948510. https://doi.org/10.1155/2013/948510

[30] Manouras, T. and Argitis, P., 2020. High sensitivity resists for EUV lithography: a review of material design strategies and performance results. *Nanomaterials*, 10(8), p. 1593.

[31] Dhawan, A., Du, Y., Batchelor, D., Wang, H.N., Leonard, D., Misra, V., Ozturk, M., Gerhold, M.D. and Vo-Dinh, T., 2011. Hybrid top-down and bottom-up fabrication approach for wafer-scale plasmonic nanoplatforms. *Small (Weinheim an der Bergstrasse, Germany)*, 7(6), p. 727.

[32] Shen, G. and Chen, D., 2009. One-dimensional nanostructures and devices of II–V group semiconductors. *Nanoscale Research Letters*, 4, pp. 779–788.

[33] Kumar, S., Bhushan, P. and Bhattacharya, S., 2018. Fabrication of nanostructures with bottom-up approach and their utility in diagnostics, therapeutics, and others. In: Bhattacharya, S., Agarwal, A., Chanda, N., Pandey, A., and Sen, A. (eds.), *Environmental, Chemical and Medical Sensors. Energy, Environment, and Sustainability*. Springer, Singapore. https://doi.org/10.1007/978-981-10-7751-7_8

[34] Freyhardt, H.C. and Ploog, K., 1980. Molecular beam epitaxy of III–V compounds. In III–V Semiconductors (pp. 73–162). Springer Berlin Heidelberg.

[35] Cho, A.Y. and Arthur, J.R., 1975. Molecular beam epitaxy. *Progress in Solid State Chemistry*, 10, pp. 157–191.

[36] Rothemund, P.W., 2005, November. Design of DNA origami. In *ICCAD-2005. IEEE/ACM International Conference on Computer-Aided Design, 2005* (pp. 471–478). IEEE.

[37] Whitesides, G.M., Mathias, J.P., and Seto, C.T. (1991). Molecular self-assembly and nanochemistry: a chemical strategy for the synthesis of nanostructures. *Science* 254 (5036), pp. 1312–1319.

[38] Hench, L.L. and West, J.K., 1990. The sol-gel process. *Chemical Reviews*, 90 (1), pp. 33–72.

[39] Wang, X., Li, Z., Shi, J. and Yu, Y., 2014. One-dimensional titanium dioxide nanomaterials: nanowires, nanorods, and nanobelts. *Chemical Reviews*, 114 (19), pp. 9346–9384.

[40] Wang, J.X., Sun, X.W., Yang, Y. and Wu, C.M.L., 2009. N–P transition sensing behaviors of ZnO nanotubes exposed to NO_2 gas. *Nanotechnology*, 20 (46), p. 465501.

[41] Choi, K.S. and Chang, S.P., 2018. Effect of structure morphologies on hydrogen gas sensing by ZnO nanotubes. *Materials Letters*, 230, pp. 48–52.

[42] Sen, R., Govindaraj, A. and Rao, C.N. (1997). Carbon nanotubes by the metallocene route. *Chem. Phys. Lett.* 267 (3–4): 276–280.

[43] Soares, O.S.G.P., Rocha, R.P., Gonçalves, A.G., Figueiredo, J.L., Órfão, J.J.M. and Pereira, M.F.R., 2015. Easy method to prepare N-doped carbon nanotubes by ball milling. *Carbon*, 91, pp. 114–121.

[44] Popov, V.N., 2004. Carbon nanotubes: properties and application. *Materials Science and Engineering: R: Reports*, 43 (3), pp. 61–102.

[45] Singh, N., Buddharaju, K.D., Manhas, S.K., Agarwal, A., Rustagi, S.C., Lo, G.Q., Balasubramanian, N. and Kwong, D.L., 2008. Si, SiGe nanowire devices by top–down technology and their applications. *IEEE Transactions on Electron Devices*, 55 (11), pp. 3107–3118.

[46] Hobbs, R.G., Petkov, N. and Holmes, J.D., 2012. Semiconductor nanowire fabrication by bottom-up and top-down paradigms. *Chemistry of Materials*, 24 (11), pp. 1975–1991.

[47] Shakthivel, D., Ahmad, M., Alenezi, M.R., Dahiya, R. and Silva, S.R.P., 2019. *1D Semiconducting Nanostructures for Flexible and Large-Area Electronics: Growth Mechanisms and Suitability*. Cambridge University Press.

[48] Wu, Y., Yan, H., Huang, M., Messer, B., Song, J.H. and Yang, P., 2002. Inorganic semiconductor nanowires: rational growth, assembly, and novel properties. *Chemistry–A European Journal*, 8 (6), pp. 1260–1268.

[49] Wagner, R.S. and Ooherty, C.J., 1968. Mechanism of branching and kinking during VLS crystal growth. *Journal of the Electrochemical Society*, 115 (1), p. 93.

[50] Ghassan, A.A., Mijan, N.A. and Taufiq-Yap, Y.H., 2019. Nanomaterials: an overview of nanorods synthesis and optimization. *Nanorods and Nanocomposites*, 11(11), pp. 8–33.

[51] Li, L.S. and Alivisatos, A.P., 2003. Origin and scaling of the permanent dipole moment in CdSe nanorods. *Physical Review Letters*, 90 (9), p. 097402.

[52] Liu, J., Tanaka, T., Sivula, K., Alivisatos, A.P. and Fréchet, J.M., 2004. Employing end-functional polythiophene to control the morphology

of nanocrystal-polymer composites in hybrid solar cells. *Journal of the American Chemical Society*, 126 (21), pp. 6550–6551.

[53] Mutiso, R.M., Sherrott, M.C., Rathmell, A.R. et al. (2013). Integrating simulations and experiments to predict sheet resistance and optical transmittance in nanowire films for transparent conductors. *ACS Nano* 7 (9), pp. 7654–7663.

[54] Kim, J., Campbell, A.S., de Ávila, B.E.F. and Wang, J., 2019. Wearable biosensors for healthcare monitoring. *Nature Biotechnology*, 37 (4), pp. 389–406.

[55] Lee, H., Hong, Y.J., Baik, S., Hyeon, T. and Kim, D.H., 2018. Enzyme-based glucose sensor: from invasive to wearable device. *Advanced Healthcare Materials*, 7 (8), p. 1701150.

[56] Shin, H., Seo, H., Chung, W.G., Joo, B.J., Jang, J. and Park, J.U., 2021. Recent progress on wearable point-of-care devices for ocular systems. *Lab on a Chip*, 21 (7), pp. 1269–1286.

[57] Tang, L., Chang, S.J., Chen, C.J. and Liu, J.T., 2020. Non-invasive blood glucose monitoring technology: a review. *Sensors*, 20 (23), p. 6925.

[58] Feng, S., Farha, F., Li, Q., Wan, Y., Xu, Y., Zhang, T. and Ning, H., 2019. Review on smart gas sensing technology. *Sensors*, 19 (17), p. 3760.

[59] Kim, E., Lee, S., Kim, J.H., Kim, C., Byun, Y.T., Kim, H.S. and Lee, T., 2012. Pattern recognition for selective odor detection with gas sensor arrays. *Sensors*, 12 (12), pp. 16262–16273.

[60] Moon, J., Park, J.A., Lee, S.J., Zyung, T. and Kim, I.D., 2010. Pd-doped TiO2 nanofiber networks for gas sensor applications. *Sensors and Actuators B: Chemical*, 149 (1), pp. 301–305.

[61] Feng, C., Kou, X., Liao, X., Sun, Y. and Lu, G., 2017. One-dimensional Cr-doped NiO nanostructures serving as a highly sensitive gas sensor for trace xylene detection. *RSC Advances*, 7 (65), pp. 41105–41110.

[62] Jo, S., Sung, D., Kim, S. and Koo, J., 2021. A review of wearable biosensors for sweat analysis. *Biomedical Engineering Letters*, 11 (2), pp. 117–129.

[63] Kang, S.G., Song, M.S., Kim, J.W., Lee, J.W. and Kim, J., 2021. Near-field communication in biomedical applications. *Sensors*, 21 (3), p. 703.

[64] Kaisti, M., Boeva, Z., Koskinen, J., Nieminen, S., Bobacka, J. and Levon, K., 2016. Hand-held transistor-based electrical and multiplexed chemical sensing system. *ACS Sensors*, 1 (12), pp. 1423–1431.

[65] Baker, L.B., Model, J.B., Barnes, K.A., Anderson, M.L., Lee, S.P., Lee, K.A., Brown, S.D., Reimel, A.J., Roberts, T.J., Nuccio, R.P. and Bonsignore, J.L., 2020. Skin-interfaced microfluidic system with personalized sweating rate and sweat chloride analytics for sports science applications. *Science Advances*, 6 (50), eabe3929.

[66] Velasco, G. and Schnell, J.P., 1983. Gas sensors and their applications in the automotive industry. *Journal of Physics E: Scientific Instruments*, 16 (10), p. 973.

[67] Leelakumar, M., 2020. Design of Electronic Control for Diesel Engines. *Design and Development of Heavy Duty Diesel Engines: A Handbook.* Springer, Singapore, pp. 795–830.

[68] Masikini, M., Chowdhury, M. and Nemraoui, O., 2020. Metal oxides: Application in exhaled breath acetone chemiresistive sensors. Journal of The Electrochemical Society, 167(3), p. 037537.

[69] Raza, M.H., Kaur, N., Comini, E. and Pinna, N., 2020. Toward optimized radial modulation of the space-charge region in one-dimensional SnO_2–NiO core–shell nanowires for hydrogen sensing. *ACS Applied Materials & Interfaces*, 12 (4), pp. 4594–4606.

[70] Nikfarjam, A., Hosseini, S. and Salehifar, N., 2017. Fabrication of a highly sensitive single aligned TiO2 and gold nanoparticle embedded TiO2 nano-fiber gas sensor. *ACS Applied Materials & Interfaces*, 9 (18), pp. 15662–15671.

[71] Lloyd Spetz, A., Unéus, L., Svenningstorp, H., Tobias, P., Ekedahl, L.G., Larsson, O., Göras, A., Savage, S., Harris, C., Mårtensson, P. and Wigren, R., 2001. SiC based field effect gas sensors for industrial applications. *Physica Status Solidi (a)*, 185 (1), pp. 15–25.

[72] Moon, S.E., Choi, N.J., Lee, H.K., Lee, J. and Yang, W.S., 2013. Semiconductor-type MEMS gas sensor for real-time environmental monitoring applications. *Etri Journal*, 35 (4), pp. 617–624.

[73] Ibrahim, R.K., Hayyan, M., AlSaadi, M.A., Hayyan, A. and Ibrahim, S., 2016. Environmental application of nanotechnology: air, soil, and water. *Environmental Science and Pollution Research*, 23, pp. 13754–13788.

[74] Yamazoe, N. and Miura, N., 1992. New approaches in the design of gas sensors. In *Gas Sensors: Principles, Operation and Developments* (pp. 1–42). Dordrecht: Springer Netherlands.

[75] Jie, J., Zhang, W., Bello, I., Lee, C.S. and Lee, S.T., 2010. One-dimensional II–VI nanostructures: synthesis, properties and optoelectronic applications. *Nano Today*, 5 (4), pp. 313–336.

[76] Chen, Q., De Marco, N., Yang, Y.M., Song, T.B., Chen, C.C., Zhao, H., Hong, Z., Zhou, H. and Yang, Y., 2015. Under the spotlight: The organic–inorganic hybrid halide perovskite for optoelectronic applications. *Nano Today*, 10 (3), pp. 355–396.

[77] Luo, L.B., Xie, W.J., Zou, Y.F., Yu, Y.Q., Liang, F.X., Huang, Z.J. and Zhou, K.Y., 2015. Surface plasmon propelled high-performance CdSe nanoribbons photodetector. *Optics Express*, 23 (10), pp. 12979–12988.

[78] Dou, L., Yang, Y., You, J., Hong, Z., Chang, W.H., Li, G. and Yang, Y., 2014. Solution-processed hybrid perovskite photodetectors with high detectivity. *Nature Communications*, 5 (1), p. 5404.
[79] Afzaal, M. and O'Brien, P., 2006. Recent developments in II–VI and III–VI semiconductors and their applications in solar cells. *Journal of Materials Chemistry*, 16 (17), pp. 1597–1602.
[80] Tvingstedt, K., Malinkiewicz, O., Baumann, A., Deibel, C., Snaith, H.J., Dyakonov, V. and Bolink, H.J., 2014. Radiative efficiency of lead iodide-based perovskite solar cells. *Scientific Reports*, 4 (1), p. 6071.
[81] Duan, X., Huang, Y., Cui, Y., Wang, J. and Lieber, C.M., 2001. Indium phosphide nanowires as building blocks for nanoscale electronic and optoelectronic devices. *Nature*, 409 (6816), pp. 66–69.
[82] Huang, Y., Duan, X. and Lieber, C.M. (2005). Nanowires for integrated multicolor nanophotonics. *Small* 1 (1), pp. 142–147.

Index

A

Actuators
- overview and applications in IoT, 2
- in smart devices, 21

Amperometry; *see also* Conductimetry; Potentiometry
- applications in healthcare as biosensor, 126
- definition and principles, 8

Analog-to-digital converter (ADC)
- interfacing with IoT devices, 2–3
- use in sensor systems, 2

Antibodies
- in biosensors, 10, 124
- role in immunosensors, 167–168

B

Biosensors; *see also* Optical biosensors; Piezoelectric biosensors
- applications of biosensor, 134–139, 152
- based on metal oxide semiconductor nanostructures, 121–134
- definition and principles, 3–4, 123–124
- electrochemical biosensors, 5–9, 125
- FET-based biosensors, 11, 55
- role in human health monitoring, 155–159
- role in IoT, 172

Biowarfare agents (BWA)
- detection using biosensors, 137

Blood pressure sensors
- applications in healthcare, 172

C

Cancer detection
- biosensors for, 150–165
- electrochemical approaches, 153–155
- role of nanomaterials, 155

Carbon nanotubes (CNTs)
- applications in gas sensors, 71
- definition, 131
- in nanomaterial-based sensor, 170
- properties for sensor development, 63, 66
- use in biosensors, 154–158

Charge transport
- in conjugated polymer, 45–48
- in non-conjugated polymer, 48–52

Chemical vapor deposition (CVD)
- in bottom-up approach, 169
- in sensor fabrication, 19–21, 74–77

Conductimetry
- applications in biosensors, 126
- definition, 8–9
- gas detection, 67

Cross-sensitivity
- methods to reduce, 72

D

Data storage in IoT
- sensor networks role, 4–5

Diagnostics
- biosensor role in, 136–137, 156
- for cancer detection, 157

DNA-based sensors

applications in healthcare, 137–138, 156–158
in electronic devices, 11–12

E

Electrochemical biosensors
advantages over other biosensors, 125
applications in healthcare, 126
for cancer detection, 131
overview and principles, 124–125
with based on metal oxide semiconductor nanostructures, 127–128
Electrochemical impedance spectroscopy (EIS); *see also* Immunosensors
applications in biosensors, 135–136
overview, 9
Enzyme-based sensors (EnFET)
for glucose detection, 10–11

F

Field-effect transistor (FET)
based on nanomaterials for gas sensor fabrication, 73–101
biosensing applications, 10–12, 52–55
polymer semiconductors, 52–58
Fluorescent biosensors
overview, advantages and applications, 136

G

Gas sensors
applications, 173–174
based on metal oxide semiconductors, 59–78
core-shell based, 99–100
effect of oxygen vacancy, 89
enhancement of gas sensor properties, 81–87
fabrication techniques, 74–80
with novel nanostructures, 90–97
working principles, 61–65
Glancing angle deposition (GLAD)
mechanism, 21–25
synthesis of metal oxide NWs in sensor fabrication, 25–28

H

Hybrid nanostructures
advantages and applications, 134–139
role in IoT, 3–5
as solid-state sensors, 1–15, 127–134

I

Immunosensors
applications in diagnostics, 151–152
applications in environmental monitoring, 137–138
based on CNTs, 131
based on EIS, 9
Impedance-based sensors; *see also* Potentiometry; Electrochemical impedance spectroscopy (EIS)
electrochemical techniques, 125–126
for health monitoring, 135–136
Internet of Things (IoT)
architecture, 4
integration with nanostructures, 3–5
sensors and applications, 1–4, 123, 134–139

L

Liquid phase epitaxy (LPE)
for nanowire sensor fabrication, 74

M

Machine learning
role in sensor data analysis, 154
Metal oxide semiconductors
in biosensors, 127–134, 155
in gas sensors, 61
overview, 16–19, 59–61
various nanostructure synthesis for sensors, 19–35, 75
Microfluidic sensors
for diagnostics, 135–137

N

Nanomaterials
application as IoT-enabled sensors, 172–179
in biosensing applications, 155–159

different synthesis techniques, 168–172
as metal-oxide semiconductor, 127–128
role in FET sensors, 10
Nanoparticles (NPs)
applications in gas sensors, 77–79
for biosensing, 132–134
Nanowires (NWs)
in biosensor, 130–131
in gas sensor, 75–80, 99
in IoT-enabled sensor, 171
synthesis by GLAD, VLS and sol-gel technique, 16–35

O

Optical biosensors
advantages and applications, 134–139
definition and working, 126
for real-time monitoring, 152

P

Piezoelectric biosensors, 127
role of nanoparticles, 132
Point-of-care testing (POCT); *see also* Internet of Things (IoT)
overview, 1–15
with IoT-enabled sensors, 172–173
Polymer semiconductors
in FET sensors, 10, 39–58
Potentiometry, 6

R

Reversibility (sensors)
in biosensing applications, 133–134
of sensor, 5

S

Selectivity (sensors)
in biosensor, 151–153
in gas sensor, 67–72
of sensors, 5
Sensitivity (sensors); *see also* Selectivity; Reversibility
in biosensor, 124, 150–165
in gas sensor, 66–71
in IoT-enabled sensor, 166–185
in metal-oxide based nanostructure, 127–134
of sensors, 5
Smartphones
role of IoT sensors, 1–3
role of nanomaterials, 153–154
Sol-Gel Technique, 33–35

T

Threshold voltage
role in FET sensors, 53–54
Transition metal dichalcogenides (TMDs)
in biosensor, 156

U

Ultraviolet (UV) sensors
as photodetectors, 17–18, 175
Urinary tract infection (UTI) sensors, 135

V

Vapour-Liquid-Solid (VLS) deposition
for metal oxide NWs synthesis, 29, 74–75

For Product Safety Concerns and Information please contact our EU representative GPSR@taylorandfrancis.com
Taylor & Francis Verlag GmbH, Kaufingerstraße 24, 80331 München, Germany

www.ingramcontent.com/pod-product-compliance
Lightning Source LLC
LaVergne TN
LVHW012329100826
845148LV00017B/542

* 9 7 8 1 0 3 2 7 3 4 4 3 9 *